ON THE SHORES OF ENDLESS WORLDS

The Search for Cosmic Life

Andrew Tomas

SOUVENIR PRESS

First published 1974 by Souvenir Press Ltd,
95 Mortimer Street, London W1N 8HP
and simultaneously in Canada by
J. M. Dent & Sons (Canada) Ltd,
Ontario, Canada

ISBN 0 285 62131 9

Filmset in Photon Times 12 on 14 pt. by
Richard Clay (The Chaucer Press), Ltd, Bungay, Suffolk
Printed in Great Britain by
Fletcher & Son, Ltd, Norwich

To

NICOLAUS COPERNICUS

(1473–1543)

on his 500th anniversary for his courage in dislodging the Earth, and man along with it, from the centre of the universe.

Acknowledgments

I gratefully acknowledge permission to reproduce illustrations granted by the following parties:

NASA, Houston, Texas, and United States Information Service, London, for the photographs of an astronaut's footprint on the moon and Pioneer 10 plaque.

Upjohn International, Inc., Kalamazoo, Michigan, for the photograph of a cell model.

Colin Richards, Photographer, Cape Town, South Africa, for the photograph of Welwitschia plant.

The London Dolphinarium for the photograph of a performance by their dolphins.

John Fairfax and Sons (Australia) Ltd for the photograph of Koko, the chimpanzee.

The Whitworth Art Gallery (University of Manchester) for William Blake's *The Angels appearing to the Shepherds.*

My thanks are due to Design Practitioners Limited of Sevenoaks, Kent, for the plates depicting insects, prehistoric man with Apollo and the explosion of a planet.

I am also thankful to my friends in Paris—Elaine Ackerman and Anne Croser for checking the manuscript.

Contents

Introduction

A starfish slowly crawled on the sea-bed. Large and small fishes darted in all directions while the seaweeds swayed to and fro. The commentator of this documentary of marine life then asked: 'Have you guessed what this sea is? The Red Sea or the Caribbean, perhaps? Or is it the Pacific? No, ladies and gentlemen, you are all wrong—we are under the ice of the Arctic Ocean.' This disclosure that life could exist near the North Pole left a mark on my mind for many years to come.

Is life an inherent property of matter or an exceptional phenomenon in the universe? Is man an evolutionary dead-end beyond which there is a vacuum? These questions are not devoid of pragmatic value because most of the problems confronting man today spring from his inability to orient himself correctly in the grand structure of Space–Time.

The theme of the panorama of life in the universe is so vast that we hope the author will be forgiven for merely sketching its outlines. This topic can be approached from many angles—among the major ones are those of biology, astronomy and philosophy. In a volume of this format it is hardly possible to cover even one of these fields thoroughly. Hence, at the risk of being accused of dilettantism, the different areas of research have merely been outlined. Only by this technique can an overall picture of life in the universe be presented.

Generally speaking, the dialectical materialists of ancient Greece have provided the most satisfactory method of dealing with the problem. But in those areas they have not tackled, we will wander into the region of metaphysics in search of some answers.

A proper philosophical approach must be all-embracing because we live in an infinite universe. All knowledge is relative depending on the position in Space–Time. The wider the diapason of thought, the greater the possibility of discerning truth.

The book is divided into two different parts. Part I, THE TREE OF LIFE, deals with the growth of life in a vertical direction—from the infinitely little to the infinitely great. This is the road of cosmic evolution from atom to galaxy. Part II, ON THE SHORES OF ENDLESS WORLDS, is devoted to the exploration of the universe in search of signs of sentient life in boundless space. This problem could never be worked out before life has been traced to its source—matter, and then its ascent followed to higher evolutionary forms.

You will thus hear insects speak, agree that birds and animals are people, and then beyond the figures of men behold the dazzling faces of gods in the sky.

Man has a place in the framework of creation and he must know where he stands. Our mental attitudes are still suffering from the fallacy of the exaggerated significance of man that has dominated mankind for centuries.

This Space Age is an age of wonders—the earthman has walked on the moon, his mechanical eyes can see the strange terrain of Mars and his radio ears can hear echoes from remote stars.

Will man of this earth ever land on other inhabited worlds? Have extraterrestrial astronauts visited the Earth and left artefacts in the past? Is it possible to tune in on a radio station in another stellar world? These questions no longer belong to science fiction but to science proper.

Lastly, a certain clarification is called for: this work has been written from the point of view of a sentient being from another world in space. That is why some remarks may sound uncivil, conclusions strange and ideas unearthly. But each has been set forth with the best of intentions.

PART I THE TREE OF LIFE

1: *The Great Dawn*

The planet shone brightly but darkness reigned under its thick clouds. A boundless ocean of fresh water, smelling of ammonia, was spread over the planet's surface. The sea was boiling from the fiery torrents of lava thrown out by exploding volcanoes. It had rained for millions of years and the steamy heat was overbearing.

Global hurricanes and earthquakes raised mountain-high waves in the ocean. However, these storms provided this world of gloom with light when lightning rent the black clouds asunder. Now and then tornadoes bared the sky through which poured ultraviolet rays from the golden star around which the planet whirled. Then eternal night returned, to be interrupted only by the red flashes of erupting volcanoes and blue-white blazes of electric discharges in the clouds. This is not a picture of an unknown planet from a science fiction story but a scene of Earth in its infancy about 3,000 million years ago. What looked like doomsday was the genesis of life on Earth.

The cataclysmic environment of that distant period did not give the slightest hint that six, four or two-legged creatures would ever walk on the planet's crust, or that others would swim in its oceans and fly in its air. There was nothing to indicate that the light of intelligence would ever spring forth in this inferno. Biologically, the Earth was dead.

When this chemical sea of ammonia, methane, hydrogen and steam was bombarded by electric lightning and solar ultraviolet rays, a miracle took place—the oceans turned deep red, the colour of blood. It was the sign that amino-acids, the basis of life-giving proteins, had germinated. The Earth bore life helped by the fiery darts of Jupiter and the dazzling light of Apollo.

When the first cell was born in this maelstrom, its nucleus must have contained the seed of plant, fish, reptile, bird, mammal and man for here was the inception of all planetary life.

Millions of years have passed since that day of the Great Dawn. Today the whole planet is alive, even its soil is a living organism, the atmosphere teems with bacterial life and the oceans sustain myriads of marine life-forms. The flame of life burns on the North Pole as much as in the Sahara.

This genesis was duplicated in 1953 by Stanley Lloyd Miller of the University of Chicago, on the initiative of Professor Harold Urey. He made a mixture of ammonia, methane and hydrogen in a glass vessel, in another he boiled water and passed its steam up a tube into the gas mixture. Then the gases were pushed up by steam through another cooled tube back into the boiling water. Then Miller bombarded the mixture with electric sparks imitating lightning in this re-enactment of the scene of creation which, as in the Bible, supposedly took only one week. The colourless water and gases began to turn pink and then red. A chemical analysis showed that glycine and alanine were present—these are the simple amino-acids which enter into the composition of life-giving proteins.

In the conditions of Earth in its primitive stages, all the gases with which Miller experimented, existed in large quantities as did lightning bolts and ultraviolet radiation from the sun. Many other tests have been conducted and they all confirm Dr Miller's results. If amino-acids (protein-building blocks) can be created in a laboratory, then the appearance of life is a natural phenomenon in the evolution of matter. It is neither a miracle nor an accident. If the right chemical elements, temperature, ultraviolet radiation and electric sparks are supplied in proper proportion, life can be formed anywhere.

That life emerged on this planet in the manner described

here, is far from being an hypothesis. J. William Schopf of Harvard University has actually detected traces of twenty-two different amino-acids in rocks. As they were 3,000 million years old, this established the date of the Great Dawn.

Over fifty years ago the British biochemist John Haldane was the first to suggest that oxygen in the atmosphere was produced by plants. Haldane claimed there was a time when the Earth did not have any oxygen and the air was composed mainly of nitrogen and carbon dioxide. This gave science the first clue as to the type of chemical elements which existed in Earth's childhood.

Alexander Oparin, a Soviet biologist, wrote in the 1930s that life had been generated out of gases and water millions of years ago. Haldane and Oparin formed logical working theories without actually proving them.

It was Melvin Calvin of the University of California who in 1950 enlarged on Haldane's ideas by adding some of his own, namely, that several milliards of years ago there was twice as much radioactivity as there is today. So he experimented with carbon dioxide and water vapour, bombarding them with radioactivity. The result was formaldehyde and formic acid which are simple organic molecules.

Oparin's conclusions helped Stanley Miller to prove in his laboratory that amino-acids, the basis of organic life, could be created artificially. Other scientists have repeated the tests with variations, such as Abelson, Groth, von Weyssenhoff, Oro, Ponnamperuma, Fox, Pavlovskaya, Pasynski and others.

One cannot but agree with Friedrich Engels's words written a century ago that 'Life does not appear arbitrarily nor exist eternally but it appears in the process of the evolution of matter whenever the conditions are favourable'. It is obvious that as long as the building blocks of life are provided, organic forms will come. This *will to live* seems to be inherent within the core of matter itself.

The British biologist J. D. Bernal says: 'There is no such thing as the origin of life—life is originating all the time.' While it is noon for us in the day of life, a dawn may be rising somewhere else in the universe. Another solar system does not necessarily have to duplicate the terrestrial genesis of life but it may introduce variations according to the raw materials and energies available.

2: *The Flame of Life*

'You and the stone are one,' said the poet Kahlil Gibran. This is not only a metaphor but a statement of fact. Long before there could have been an organic evolution there must have been a chemical evolution of an immensely longer duration. Life on this planet was already in sight when the solar system had begun to crystallise.

From atom to molecule, from molecule to cell, from cell to tissue, from tissue to organ, from organ to organism—such is the road that leads to life. Life has been potentially locked in matter for aeons, awaiting a moment when, like a seed, it could sprout, grow and give fruit. The roots of the Tree of Life are hidden in the atomic micro-world and its branches embrace the stars and galaxies. If life once made itself, it will do it again and again. Life is being born all the time and as one system dies another appears in its stead.

Modern biology has demonstrated that there is no essential difference between the animate and inanimate. Living matter has a highly complex structure while non-living has a simpler system. An approximate guide to the identification of the living can be found in the presence of the following processes—growth, movement, metabolism, reproduction, irritability, feeding, respiration and excretion. This rule applies to plant and animal life alike.

Sir Jagadis Chandra Bose has shown that even metals and crystals possess sensitivity. Crystals, when squeezed and pressed, produce piezoelectricity in a sort of a reflex action. It is this very property that is being utilised in transistors. Actually, the struggle for life can be seen in the formation and growth of crystals and it is as dynamic as the fight for survival of animate forms.

In the atomic underworld the non-living and living seem alike because the same elements enter their composition. This is a highly important fact in understanding life.

Welwitschia mirabilis of South-West Africa deserts lives to an age of 2,000 years, proving the dynamic power of life.

Water seeks the lowest level. Life also fills every void. Nowhere is the struggle for life as dramatic as in the waterless desert. The *Welwitschia mirabilis* with its long leatherly leaves and 18-metre roots grows in the Namib Desert in South West Africa where the rainfall is but $2\frac{1}{2}$ centimetres yearly. According to Professor C. H. Bornman of the University of Natal 'there are living specimens aged approximately two thousand years and more'.[1] One can only admire the tenacity of a plant in a hostile environment which has outlived the Roman Empire!

Living bacteria have been found 29 metres beneath the ice of the South Pole. *Spherella nivalis*, a water plant, grows on snow in the Arctic. In Queen Maud Land in Antarctica Soviet scientists have discovered tiny insects which live on moss. These are of great interest to biology and also to exobiology because of the extremely low temperatures prevailing near the South Pole.[2] The spores of bacteria and fungi have been discovered in the stratosphere at an altitude of 33 kilometres. One particular type called *flavobactine* can not even live when brought down to the Earth's surface! Neither can *anaerobes* tolerate any air. Micro-organisms breed happily in the fuel tanks of jet aircraft using kerosene for food. Certain living organisms even manage to live in a deadly poison such as potassium cyanide. The radiation of the Oak Ridge atomic reactor (U.S.A.) could not kill *pseudomonas* bacteria, as these continued to grow and multiply after a lethal dose of radioactivity.

In the depths of the Indian Ocean devoid of practically all oxygen, no fish can live; however, a certain tiny octopus swims there by the millions. At the bottom of the Pacific Ocean there are thousands of forms of life in spite of the fact that the pressure there is around one thousand atmospheres, which would crush man into a biological pulp. At the level of 11 kilometres in the Pacific Ocean the American naturalist Jean Piccard has watched a living fish through the porthole of his bathyscape. Fossil bacteria from a salt mine were re-animated by the German scientist G. Dombrowsky in 1962 after 600 million years.[3] A seed of a lotus or a grain of wheat from an

[1] *Report from South Africa* (London), May, 1971, and *Lantern* (University of Natal), March, 1971.

[2] B. P. Constantinov, *Naselenny Kosmos* (Inhabited Cosmos), Moscow, 1972.

[3] *Technica Molodezhi* (U.S.S.R.), No. 1, 1964.

ancient Egyptian tomb can germinate after thousands of years. The words of Krishna in the *Bhagavat Gita* declare a scientific truth:

I say to thee—weapons reach not the Life,
Flame burns it not, waters can not overwhelm,
Nor dry winds wither it.

Life is not a specifically terrestrial phenomenon but a cosmic one. Almost all of the elements of the periodic table of Mendeleyev enter into the composition of living forms, yet these very elements are scattered widely in infinite space. The genesis of life is due to the presence of water, carbon, nitrogen and other elements. It is triggered by starshine and electricity both of which are in abundance in the universe.

The importation of bacterial life by meteorites is not entirely ruled out. The planting of the seeds of life, consciously or unconsciously, by another galactic civilisation is not impossible either. Let us recollect the rigid disinfection of the Apollo lunar modules in order not to contaminate the moon with our bacteria. However, these suppositions alone do not explain the original source of life.

We know that the basis of all organic life is the cell—that tiny but complex unit. The human cell has forty-six chromosomes, micro-bodies carrying DNA chains memorising and programming all cellular activity. This selective and regulative control usually ensures the correct reproduction of cellular patterns. The human egg and the amoeba have practically the same form and apparent structure. The essential difference between the two cells lies in their nuclei. The DNA is a blueprint of heredity, the unit of inherited information as well as a lengthy code of instructions to be obeyed by the cell. It would seem that the cell nucleus is like a brain which *knows* its purpose. The instructions to make amino-acids into proteins are genetically coded to all life forms in the same cipher. The instructions for a virus in DNA language would occupy half this book, whereas those for a human would necessitate a library of perhaps 1,500 books like this one.

Evolution elaborates cellular material into the sophisticated life forms around us. It is a process whereby a simple thing develops by stages into something more complicated. Evolution works by mutations in genes due to atomic rearrangements which create permanent alterations in the organisms leading either to new perfected forms or to the creation of freaks. Mutations are changes in cellular programmes which happen all of a sudden.

There have been six forward leaps in evolution during the 450 million years of vertebrate development—at first fish, then amphibia, reptiles, birds, mammals and men. From the time of its genesis on this planet 3,000 million years ago, life has been continuously renewing and remaking itself. The process of evolution reaches unceasingly towards perfection in form and function and if nothing else ensures survival.

What made that first cell in the primeval chemical soup of Earth divide itself? Were not its two parts the Adam and Eve of all life on this planet? What compelled life to burst forth? Can these questions be answered by mechanistic materialism which denies the existence of an evolutionary momentum in matter? On the other hand, dialectical philosophy can provide several answers by virtue of its main tenet of the inexhaustability of the atom. It is self-evident that the marked tendency of matter is towards complexification, improvement and immortality rather than towards simplification and dissipation.

On this planet life began as a one-celled protozoa. The birthplace of evolution was the sea, where sponges, jellyfish, shellfish, trilobites, molluscs, worms and arthropoda swam, crawled or attached themselves to the ocean floor. The arthropoda began the conquest of the land. We know their descendants today as shrimp, crabs, lobsters, spiders, scorpions, ticks and insects. After the fish came amphibians, then reptiles and birds, which were followed by mammals which appeared less than 200 million years ago. The genealogy of the horse, camel, elephant, rhinoceros, tapir, dog and other animals can be followed back through successive stages to prehistoric beasts. Man's ancestry can be traced only for about three million years. The evolutionary rate varies with each species. The *Brachiopod lingula* is exactly the same today as it was 500 million years ago. In past geological periods the Earth would have been unrecognisable to us with dinosaurs trampling the tropical forests and giant reptilian birds flying in the air. Throughout the whole course of terrestrial evolution forms have either died out or have improved and their intelligence grown. This process is still continuing.

Biologists claim that basic life chemicals have an inherent ability to form living systems. This disposes of the necessity for an outside agency in creation, but does not explain the rational patterns and tendencies on cellular, chromosomal and molecular levels. This problem will be discussed in another chapter.

3: *The Infinitesimal Computer*

When Darwin said, 'Talk about the origin of life, you might just as well talk about the origin of the elements,' he demonstrated his deep understanding of the source of life as cells are made of elements and elements are combinations of atoms. An atom is more complex than an electron, a molecule is more complicated in structure than an atom, and a living cell is extremely intricate in organisation compared with a molecule.

Geometrical patterns in the animate and inanimate worlds suggest that Nature is a mathematician, since the shape of a leaf, flower, shell, crystal or an animal can be expressed by a mathematical formula. Galileo once remarked that the book of Nature is written 'in the language of mathematics and its letters are triangles, circles and other geometric figures'. Nature's departments of mathematics, physics, chemistry and biology seem to be working in full co-operation as if in accordance with some master plan.

Is there a universal law by which Nature creates forms? Throughout the animate and inanimate domains definite patterns of rhythm and periodicity are found. A spiral seashell is built on the same principle as a spiral galaxy. Size is irrelevant in so far as geometrical form is concerned. Dr Hans Jenny of Switzerland has successfully shown how true this is by his *cymatics*. By making liquids, sand, metal filings and powder vibrate by means of a crystal oscillator, he obtained distinct figures reminiscent not only of a honey-comb, coral, shell but also of moon craters and spiral galaxies.[1] Obviously vibration,

[1] UNESCO *Courier*, December, 1969.

The cell, the building block of all life, looks like a world in space.

oscillation and pulsation can produce identical forms whether in the micro-world or in the infinitely vast. This denotes a universal law in creation that operates according to a mathematical blueprint. It is more than likely that the law works from the infinitely small outward. It is difficult to understand how Nature carries out her sophisticated programme unless the atom is a sort of computer—the most infinitesimal.

A computer solves problems and provides answers to questions after having been fed with *input* or information and instructions. In an *Analog computer* the input information varies with time. The *output* depends upon with what the machine has been programmed. A computer fed with data, or numerous facts for future reference, is able to remember by virtue of its memory bank.

Atoms, molecules and cells must possess coded memories in a manner similar to those of a computer. This has been established at least on a cellular level which concerns only organic matter. How about the inorganic? Let us take the example of crystals: a damaged crystal immediately repairs injury done to it and soon re-assumes its basic geometrical form. How does a cubic crystal of salt *know* that it is a cube unless it has a chemical memory bank? What makes a droplet of water in a temperature at freezing point or below crystallise into a six-pointed snowflake?

What a detailed image of a future plant is contained in the seed! What tremendous potential is stored in an acorn to enable it to grow into a giant oak!

A computer must have *soft ware* or programmes and experts to direct its work. Without them it is only useless *hardware.* Conversely, one is inevitably drawn to the idea that primordial mind must be incorporated in the atom. This concept is not new as the great philosopher Democritus spoke of the imprint of images on the fundamental substance of the universe.

Professor N. I. Kobozev of the U.S.S.R. has shown that events are remembered by the brain cells for many decades but he insists that this recording could not be made at cellular, molecular or even atomic levels in view of the temperature fluctuations of the human body which would wipe off everything registered in the brain. This suggests an interesting possibility that the recording is done on a subatomic plane.

Two hundred years ago Voltaire mentioned *intelligent*

atoms. More recently Teilhard de Chardin wrote: 'Every element of the universe contains, at least to an infinitesimal degree, some germ of inwardness and spontaneity, that is to say of consciousness.'[2]

Professor D. F. Lawden of New Zealand has made a conjecture that the principle of continuity in the universe is 'exemplified by the smooth gradation of forms from the fundamental particles to ourselves thus constituting a hierarchy at no level of which can a clear dividing line between the living and dead be distinguished'. The scientist alludes to the human brain and says: 'If consciousness is characteristic of this matter aggregate, by the principle of continuity it must also be a feature of every aggregate and ultimately of the fundamental particles.'[3] He insists that if this were not so, consciousness would arise discontinuously and then it would be possible to draw a line separating conscious from non-conscious matter forms, which science is not able to do. This scientific hypothesis supports the ancient belief of the East that all is alive and conscious in the infinite cosmos.

'Spirit is the ultimate sublimation of matter and matter the crystallisation of spirit,' says the *Kiu-Te*, a rare Tibetan book.

Thomas Mann explained consciousness as the desire of Nature *to listen to itself* and as *a hopeful–hopeless project of life to achieve self-knowledge.* This thought has been beautifully expressed by Shelley:

I am the eye with which the universe
Beholds itself and knows itself divine.

Life has a goal in its self-perfecting—the ultimate cognition of itself. It fills a void wherever it can find one. Similarly, intelligence spreads itself out as far as it can. The craving for

[2] T. de Chardin, *Let Me Explain*, London, 1970.

[3] *Nature*, 25 April, 1964 (p. 412).

information without which no living species can survive, is universal.

Julian Huxley wrote: 'As a result of a thousand million years of evolution, the universe is becoming conscious of itself.' This could not have happened were not intelligence developed in the course of a long planetary evolution. But the potentiality of mind was already contained in the atom. For did not Socrates say that 'We possess intelligence and that which we possess must be contained in the cause which created the world'? Eight centuries later Plotinus echoed the same conclusion in these words: 'The most irrational theory of all is that the elements without intelligence, should produce intelligence.'

4: *The Way of Intelligence*

The atomic computer of the cosmos works on the principle that the intelligent way is the most efficacious way. The knee-joint of a fly is just like your own. Why? Because it is intelligently constructed by Mother Nature.

Insects make nests and store food for winter. Birds migrate to warmer climates in autumn. Why? Because these are the rational things to do in order to stay alive. This inference might be questioned on the ground that this intelligence is purely instinctive. A study of the life of insects and birds proves that they are capable of learning. What is instinct? It is information and instructions locked in the genes which have been handed down from generation to generation. On the other hand, acquired intelligence is the sum total of information fed into the cell by experience. But whether it is instinctive or acquired, this knowledge is stored in the cellular structure, and that is the important thing.

The difference between the organic and non-organic is as nebulous as is the borderline between conscious and unconscious matter. By comparing a cell or an atom with a miniature computer it is possible to understand more clearly the mystery of life. A working hypothesis that life and mind are potentially present within matter, can help to solve many problems.

Planaria is a flatworm with a network of nerves and just the beginnings of a brain. Professor James V. McConnell of the University of Michigan as well as many other scientists have conducted interesting experiments with these worms. They were subjected to electric shocks which they did not like and wriggled every time the current passed through them. Then shortly before the shocks the experimenters warned the planarian worms of what was coming by a flash of light from a bulb. Soon the worms 'got wise' that every time the lamp flashed they could expect a nasty shock, and they started wriggling. Obviously the worms 'learned' to associate light flashes with electrical impulses. But that is not all. The scientists made a stew of these 'educated' planarians and fed it to ignorant ones.

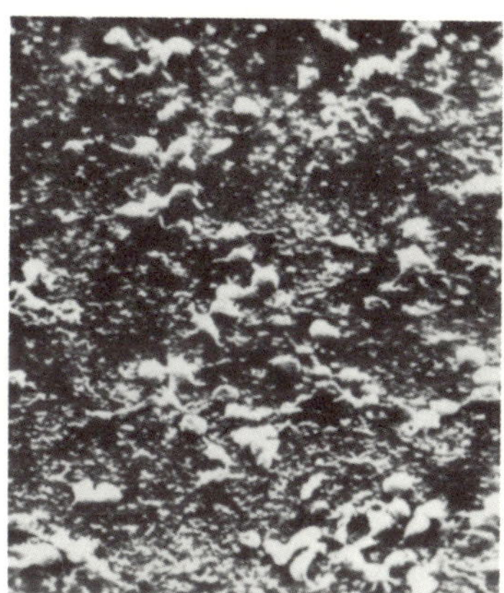

White blood cells at rest (left), and ready to attack inimical bacteria (right).

When this second group of worms was experimented with, they learned the lesson of linking the lamp light to the shocks in half the length of time it had taken the eaten group. Apparently the knowledge gained by the stewed victims was transferred to the 'cannibals' through the digestive tract instead of the genes. The moral for university students is this—chop up your professor, eat him and cut your terms in half! There is no doubt that the pedagogues will violently oppose this Lamarckian heresy.

Modern biology has discovered a semblance of intelligence on the cellular level. The DNA spirals can be compared with the memory bank of a computer having a vast information storage system. The master molecule DNA is responsible for passing on this information from one generation to the next and it is also the controller of the manufacture of proteins, the basis of organic life. The DNA code consists of four letters taken three at a time to form one genetic word, and this code is valid for all living things. The word contains the instructions for making whatever protein the body needs whenever required. These code words can demand every single amino-acid whereas others act as punctuation, signifying the completion of a message.* Even on that microscopic level there are complicated play-back mechanisms indicating when to start and when to stop carrying out the programme.

The RNA is the messenger of the DNA. After leaving its nucleus it is picked up by a ribosome (where proteins are synthesised) which reads its three-letter word. Then a transfer RNA gets the proper amino-acid from the cell fluid and delivers it. Molecular biology processes are more complex than the operation of a modern computer and they demonstrate rational, intelligent patterns on cellular and molecular levels. If these programmes were somehow nullified on all levels, life would become chaotic, coming to an end in a few hours. Finding intelligence and logic in the infinitely tiny is an exciting discovery.

The theory of pre-programmed matter does not mean that all particles of organic matter follow an overall prearranged plan. Each unit has its own programme to which it ruthlessly adheres. A virus does not grow or reproduce itself by division yet it has a power to perpetuate itself by a diabolical plan of getting into a living cell and blacking-out the coded instructions of chromosomes, thus forcing the cells to work for the invader.

Intelligence is not confined to the brain alone, for plants also possess something which is analogous to the nervous system in

* The length of DNA filaments in our bodies is truly astronomical—160,000 kilometres.

animals. With the increase and decrease of light many flowers open and close. The grapevine has spiral feelers which are sensitive enough to find the places where it can fasten its tendrils, thus directing the path of growth of the whole plant.

In fact, much evidence is now accumulating that flowers and plants experience fear, joy and pain. The emotions of plants have been studied by Dr V. M. Pushkin of the U.S.S.R. who claims that in his experiments with hypnotised subjects a plant records the feelings of the person who touches it. His conclusion is that 'Human mentality, human perception, man's thinking and memory are basically a specialisation of the information service which takes place at plant-cell level'.[1]

Cleve Backster, an American polygraph expert who trains private investigators and police to use lie-detectors, has unexpectedly discovered that plants will react to people's thoughts. He conducted experiments by which a person would picture in his mind the idea of burning a leaf, and immediately the polygraph machine recorded the plant's reaction of fear. According to Backster, his experiments have shown that plants are happy when they are watered, are afraid of violence and display anxiety upon the approach of animals.[2]

Man must feel his fundamental unity with the animal and plant kingdoms and Nature as a whole in order to understand his place in evolution. All life on this planet is the result of a cosmic process of the evolution of matter, which comprises the same elements throughout the universe.

Scientific opinion regards life as a natural process of the evolution of matter which is potentially contained within the atom awaiting a favourable environment. 'Macrocosmic evolution can only be the result of microcosmic evolution, and vice versa,' writes Louis Jacot in his work *Universal Evolution*. Evolution is not only a terrestrial phenomenon but also a

[1] United Press International, Moscow, 25 February, 1973.

[2] Richard Martin, Staff Reporter, *The Wall Street Journal*, 1972.

cosmic one as it is followed by stars and galaxies. The arrow of evolution points towards elaboration in creating superior forms. This plan was hidden in the first cell born in the chemical soup which covered the Earth 3,000 million years ago. The mystery of life-patterns lies within matter itself.

The story of evolution is nothing less than the victory of the more intelligent and adaptable. Mind *had* to appear as it is dangerous, even impossible, to live without it. Man has reached great heights because of reason, and his awareness of the world around him is immense when compared with that of the animals. The ancient Chinese scholar Hsun Ching pointed to the human mind as the source of man's supremacy: 'In strength man does not equal the ox, nor in the power of running—the horse, and yet he can utilise the ox and horse.'

In evolution there is no equality, neither good or evil, for Nature tries to develop excellence and greatness as the ultimate objective. The difference between evolving lives is mainly in the degree of intelligence and the extent of consciousness. There have been small and giant creatures. Only one factor has always remained—Reason, without which a life-form can not hope to survive for any length of time.

5: *Insect Civilisation*

In *Gulliver's Travels* Swift wrote of a voyage to the land of Lilliput inhabited by little creatures who were skilful engineers, mathematicians and mechanics. There exists a real kingdom of Lilliput where civilisation has flourished for millions of years before the appearance of man on Earth. It is the civilisation of bees, ants and termites which are more efficient than any human society, and immensely more numerous than the human race. When a complicated social organisation conceives the division of labour with the dual purpose in mind of the survival and the prosperity of a community, it deserves the name of civilisation. In this society the individual's life is totally identified with the interests of the community.

The first arch ever built on this planet was not created by man but by termites in an era when dinosaurs still roamed the world. Bees were pioneers in aerial navigation as they travelled, guided by the sun, millions of years ago.

Ants were the earliest agriculturalists. They have produced 'bread' from grain, kept 'cows' and stocked meat. The ant society consists mostly of infertile females who do what humans consider to be men's jobs as soldiers and workers. The queens are the only true females as they lay thousands of eggs a day. One peculiarity of the world of ants will surely delight the girls of the Women's Liberation movement. In mental capacity the male ants are definitely inferior to the females of this underground civilisation.

For an average ant the day starts very early with what might pass for a yawn; then comes the grooming during which the ant becomes more and more alert. Every ant has a job to do—

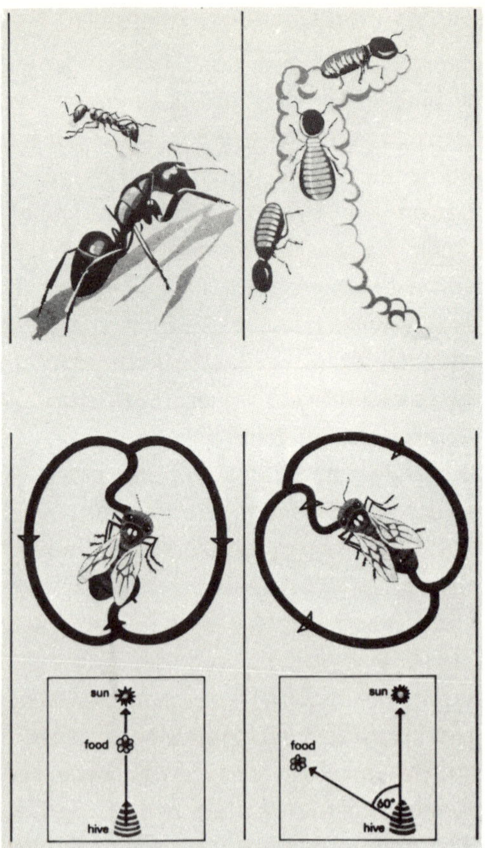

(Top left) Ants display unbelievable teamwork. (Top right) Termites built the first arch on Earth millions of years ago.
(Bottom) By her dance language the bee conveys information on the site of food supplies, its distance and type of flowers to be found.

some are soldiers defending the commune from enemies while the majority are workers, whose tasks can include anything from foraging to looking after the aphides (ant cattle), growing mushrooms, nursing eggs and grubs or cleaning the underground city.

When a scout spots a dead grasshopper it begins to run around in an agitated manner as if to say 'Can't you see I am very excited? Guess why?' When others locate the prey, news reaches the anthill and hundreds of workers run to the site, including the butchers whose duty is to cut the meat in slices for storage. The organisational work usually takes so short a time, it would be the envy of many an executive.

But there is one day in every colony on which ants do not work—it is their Royal Wedding Day. By this time the virgin queens and young males have grown wings to take them on their nuptial flight. On this warm sultry day excitement spreads throughout the countryside affecting millions of ants. In view of the large size of the young queens with their wings, doorways are enlarged to allow them to get out into the open where they assume their position for take-off, like planes. Worker ants hold them to the ground and chase away the overenthusiastic males for it is high in the air that fertilisation by the strongest male ants must take place. When the nuptial voyage is over, each queen will found a colony of her own or share it with another queen.

Soon everything is back to normal in the anthill. Aphis cows are kept in corrals by cowgirls inside the ant city. They have to be milked and taken out to pasture and brought back to the nest. The aphides' eggs are carefully looked after until they hatch out. The milking is done by gently stroking the 'cattle' with the ants' feelers. The fate of old aphides is the same as that of old cows—they are killed for meat. When the honey dew is secreted by an aphis or coccid and drops on the ground it ferments and turns into an alcoholic drink. Drunken ant cowgirls are thrown out of the community as irresponsible individuals unworthy to reside in the perfect communal society of the ants.

Ants have ten or more *scent words* by means of which they convey information. They can also draw a scent line on the ground showing the direction of food supply. In order to recognise a friend or a member of the same community by scent the ants tap each other's feelers. Social feeding and grooming are also practised by the ants.

The scouts of the multi-million army-ants of South America which destroy everything in their path, can write four signal scent words on the ground—*wait*, *advance*, *attack*, *surround*, depending upon the command. So far no satisfactory answer has been offered to explain how the 'general staff' conveys its orders so quickly in a synchronised attack with a front line stretching for a kilometre. Ants wage wars only for territory or source of food and in warfare they manifest as much intelligence, if not more, than in their work. Like nations, ants make political alliances and engage in certain strange activities which could be interpreted as insect diplomacy. Let us suppose that a serious political conflict occurred among the ants in your own backyard a few weeks ago. Without your knowledge the situation may have been degenerating for some time. Perhaps ants from the Black City met citizens of the Green Metropolis in the contested territory while both were foraging. They would merely have waved their antennae, facing each other as if 'talking'. Then negotiations stopped and the Black City scouts climbed blades of grass and bushes to have a look round and assess the situation. As no-man's-land became emptier and the appearance of scouts at observation points more frequent, the Greens started gathering in groups as if to discuss defence plans and prepare for war.

The Blacks attacked by hundreds in small skirmishes. In the meantime the main force of thousands was slowly advancing towards the Green Metropolis. The defenders put up a strong army to meet the invaders who then performed a strategic manoeuvre surrounding the enemy in a pincer movement and sealing its doom. The perfect timing of the advance of the three columns supposes that a council of war was held which planned and conducted the attack. Yet all the ants looked alike, nor were any generals visible. This raises the question—are ants intelligent group organisms or groups of intelligent organisms?

Ants take prisoners of war—eggs which when hatched will become their faithful slaves. And some victors insist on being carried by these slaves like Roman legionnaires on litters. Slavemaking ants are clever enough to capture robust yet gentle slaves as they do not want any trouble from a crawling Spartacus or his kind. However, wars among the ants are exceptions rather than the rule.

In peace-time ants are roadbuilders, engineers, herdsmen, weavers, farmers. All this requires intelligence and organisational ability, while there is no doubt that much of the behaviour of ants is instinctive yet some is guided by acquired experience. Ants memorise their trails and hunting places, and experiments with mazes have proved that some ants are cleverer than others.

Ants use two systems of astronomical orientation—they can find their way by the sun or the moon, depending on the time of day. In this regard their knowledge of astronomy may be more extensive than of some humans. I remember an occasion when, after crossing the equator, and flying over north Australia on a moonlit night, I made a curt remark to a lady—'The moon is upside down.' The passenger whispered a few words to the stewardess, apparently expressing concern for my state of mind. The stewardess's painted smile disappeared and she strolled into the pilot's cabin. In a few moments the girl and the smile reappeared: 'Yes, the captain says the moon is upside down because we are down-under.'

The harvesting ants of Europe collect grains of wheat, chew it making it a kind of bread and store it in underground granaries. Should the grain become damp because of excessive rainfall, the ants take all the grain into the sunshine immediately the rain has stopped and keep a close watch on it as it dries out. In South America ants walk with a piece of leaf over their heads. They are called parasol ants because of the way they protect their heads from the burning sun. These ants chew the leaves they have carried and use them to grow fungus. The ants cut up these mushrooms for food when they are only a few millimetres high which, if allowed their full growth, would attain 15 centimetres. The organisational problems involved in mushroom growing by the ants are very complex. There must be just the right number of workers to attend to these gardens without draining the labour force at the same time. On the other hand, too few workers would not be able to stop the uncontrolled growth of the fungus which could thus become a real jungle.

The tailor ant holds in its mandibles a weaving larva which it uses as a shuttle. It joins leaves with a silken thread emitted by the grub, out of which it builds its nest. From these examples one can see that an element of reasoning is present in the ants. It is therefore no exaggeration to speak of the ant civilisation, because they had engineers, warriors, farmers, cattle-breeders and tailors, millions of years before man. They have created a permanent society without private ownership and the ants in a commune are so united that they cannot exist individually.

In contrast to much of humanity, ants can control their population. The queen produces only the required number of workers and soldiers by glandular genetic control. We still do not know how the queens obtain their population figures if there is no continuous census taken. It is well to keep in mind that insects comprise 80 per cent of all living creatures, thus one could almost call the Earth the planet of the insects.

It is worth noting that intelligence and psyche do not necessarily correspond with the weight of the brain of insects or animals. Evidently the quantity of the brain's mass is less important than its quality.

In the eighteenth-century the mathematician Maraldi was given the following problem to solve—'With minimum material what should the form of a container be so as to give maximum capacity?' Maraldi's answer was—the hexagon. Yet the bees made this discovery millions of years before Maraldi when they built their first honeycomb cell.

The centre of the beehive is the queen, which lays up to 2,000 eggs every twenty-four hours during the warmest months. There are only a few hundred males in a colony of some 50,000 workers, and these workers are barren females as it is with the ants. After swarming, when the queen and the young queens are fertilised, the colony is broken up. The males having performed their function are chased outside to be later killed by frost. The following summer new males will hatch, to play their part and follow the same destiny.

In an organised insect society one of the main duties of the worker is to bring in food. So bees collect nectar and pollen; these are converted into honey which is stored in the honeycomb storage tanks.

A young bee begins life as a nurse looking after the eggs and grubs, then starts making wax. Other tasks await such as cleaning and ventilation, for the beehive must be kept hygienically clean. Bees even discovered a process for air-conditioning. In summer they bring water into the hive, and the fluttering of their wings evaporates it, thus reducing the temperature. In winter the bees gather in a cluster to generate warmth. It is the sugar in their food that keeps them warm and maintains a temperature of 13°C inside the hive, even when there is frost outside.

After performing a number of tasks the worker bee is then allowed to fly on food-gathering expeditions to flower fields. Towards the end of their lives ageing bees assume the duties of hive guards. Nevertheless, the occupational assignments are not rigidly fixed and it is the immediate requirements of the community that decide just what should be done and when.

A stray bee which enters a strange colony is manhandled by

the police bees at the gate. If the intruder does not fight back but remains calm, it will gradually acquire some of the scent of that beehive and so will in time become 'naturalised'. The ants are subject to these same naturalisation laws in a foreign anthill.

Bees can see polarised light and even thick clouds do not prevent them from seeing the sun's disc in ultraviolet radiation. Like a transcontinental jet navigation officer using his instruments, the bee automatically records the angle of the sun, force of wind, distance to destination and calculates the duration of flight. The bee has been able to measure the speed of its flight for millions of years, yet our aeronautics achieved this technique only a few decades ago. This does seem incredible, but many experiments with bees have confirmed these conclusions. The bees also learn to recognise the landmarks of the countryside, such as a church on a hill or a large tree. This represents acquired knowledge as opposed to instinctive behaviour. The more trips the bee makes to a certain site, the easier it is for it to remember the route.

A great discovery was made before the Second World War by the German Professor K. von Frisch concerning communication and transmission of information by the bees. When a bee has located a new flower field, it will return to the hive to describe the location of the site and the quantity of food available. Then the worker bees will detach a labour force necessary to bring the nectar home. The bee accomplishes this communication by its dance on the vertical face of the honeycomb, in the form of a figure 8. If the food is near the hive, the forager circles first to the left and then to the right. If the food is in the same direction as the sun, the informing bee runs upward on the vertical side of the beehive. If it is in the opposite direction, then it shows this by performing a downward run. The faster the run, the nearer the source. The information regarding the type of flowers discovered is given by the scent of flowers emanating from the scout. The quantity of the nectar available is shown by the degree of the forager's excitement.

If the source of food is at an angle from the sun's direction the bee estimates that angle like any good navigator, then flies back to the hive. Should the distance be several kilometres, the bee will signify the distance by the slowness of her dance. It is astonishing that the scout makes allowance for the displacement of the sun during the homeward flight just as a pilot or astronomer, and indicates the direction of the sun *at the time of the dance* and not at the moment of spotting the field. At once the bees follow the explicit instructions, place themselves on the appropriate course relative to the sun and then take off. The bee can even tell of a detour to be made if there are obstructions on the way.

The indication of distance is done in this manner: if the bee makes 10 short turns in 15 seconds, the signal means that the discovery is within a 100-metre radius from the hive, but if the scout makes only 3 turns in 15 seconds, then the distance to the field is over 3 kilometres. The bee can also show, by lifting or lowering its head, whether the site can be reached by flying towards or away from the sun. All this signifies communication of information from one member of the insect community to others and demonstrates that there is intelligence in the behaviour of the bees, fantastic though it may seem.

In a manner similar to that of human beings, bees sometimes engage in political disputes, and the majority has to decide the issue. If one group of scouts finds a good spot for a new nest, it will advertise its advantage by a symbolic dance. Now if another scout group discovers an equally suitable shelter in another locality, it will put on the same performance. This dance contest can last for hours and finally the bees choose the most convincing as winner, which they follow to the new site with their young queen.

A Soviet biologist, Dr G. A. Mazokhin-Porshniakov, claims that bees can understand mathematics and geometry. He prepared a series of cards with a sequence of round dots, then placed a small dish of syrup on the card with three dots and let the bees have a feast. A few days later he prepared a fresh set of cards so that there would be no scent from the bees' last excursion to card No. 3, and the bees all went directly to the card with the three dots, expecting a treat. Apparently they were able to count the number of dots on each card, for the scientist then put the syrup on card No. 7, and when they came the next day the bees expected the food to be placed on the card with seven dots.

Subsequently, the experimenter tried cards with geometrical figures such as a square, a triangle, a cross, a circle and even more complicated patterns. The bees had no difficulty in identifying them. During one of the tests, the professor and his assistants thought the bees had made a mistake by going to the wrong card, but on checking their notes discovered to their utmost embarrassment that it was the bees who were right after all. Dr Mazokhin-Porshniakov's natural conclusion was: 'It is hardly reasonable to continue the argument that the behaviour of insects and vertebrates is basically different.'[1]

We shall now continue our journey deeper into the Land of Lilliput. The next creatures we meet actually built the first pyramids and cathedrals about 200 million years ago. These are the termites or white ants, the latter being a name which they would hate if they understood our language, since ants are their mortal enemies. Despite their destructiveness where houses are concerned, termites do merit respect. In the course of centuries, grain by grain they have erected colossal structures, some 100 metres in diameter, weighing thousands of tons. The huge nests are exactly oriented to the north–south axis as though a compass had been used during their construction.

[1] *Nedelia*, No. 28, July, 1971 (U.S.S.R.).

If we could reduce our size to that of the termites and could somehow illuminate their nest, a fantastic sight would open to our curious gaze—a giant cave with columns, alcoves and innumerable corridors. Perhaps with a stretch of the imagination, we might even compare it with Notre Dame, St Peter's or any fabulous palace.

The queen lies in the vaulted royal chamber around which marches an endless procession. Thousands of workers feed the queen, kiss and groom her majestic body while others carry off the eggs which the queen lays at the rate of 150,000 a day. The king stands close to her, occasionally fertilising this live grub machine. Poor thing! No one kisses him, not even her majesty.

There is complete order in the termitary and the dense traffic is well regulated. The stream of termites flowing outward in search of food—timber, which they have to mince—never conflicts with the line of workers coming home.

How do termites build their vast underground cities? Are the plans of these structures preprogrammed in the genes of these insects? Professor Rémy Chauvin thinks that the intelligence of social insects comes from an *inter-connection* of their little brains and even compares a 50-thousand insect nest to a computer. He says: 'Let us suppose that the little ferrite rings of the electronic brain are provided with legs, move about and only come back to join the whole on special occasions—you would then have a machine very like a hive.'[2]

This supports our view that mind is co-extensive with matter and that on cellular and molecular levels self-programming takes place to establish rational patterns in order to ensure the continuation of the particular species.

[2] R. Chauvin, *Animal Societies*, London, 1971.

6: *Animal Kinship*

In the Lilliputian world we have just left, all creatures had a protective coat of armour, probably to compensate for the lack of skeleton. The world we are now entering is made up of much larger creatures, all of which do possess a skeleton. They can crawl on land, swim in water and fly in the air. Some of them are very clever indeed.

The arctic tern, of the seagull family, nests during the arctic summer in northern Alaska, Canada, Greenland or in the north of Europe and Asia. At this season it gets more sunlight during the white nights when the North Pole is turned towards the sun. This lover of cold sunshine then migrates to similar conditions in the Antarctic by a route stretching from 13,000 to 20,000 kilometres. When the South Pole summer is over, it flies back to the North Pole. In other words, terns find themselves spending the northern summer with the Eskimos and the southern one in Antarctica with the penguins because in this manner they can make the most of the sun for the longest time each year.

How do they do it? Vision enables them to orient themselves by the position of the sun. They have an internal chronometer to determine their position and they get their bearings by observing the length of the arc between the sun and the horizon. Our navigators had no compass until the thirteenth century while the sextant was only invented in 1730. But the birds have always been able to manage without them.

This system operates effectively only if the sky is clear, for if there are heavy clouds the terns have to rise above them. What happens at night? The terns plot their course by the stars. It may sound a little far-fetched to say that migratory birds apply astronomy in their long flights but that is a fact discovered by experimentation in planetariums in Germany. When migratory birds were released in a planetarium depicting the stars of an autumn sky, they immediately took a south-western course, towards the Mediterranean, their usual winter home. As the terns fly south they determine the positions of the principal constellations of the northern hemisphere such as the Big and Little Dippers, Cassiopeia or Draco. Flying south the birds 'look over their backs' to see where the constellations are, making corrections even for the circumpolar movement of the stars as would an astronomer. Approaching the equator they look out for the Southern Cross and the bright star Canopus. Navigation by the sun and stars does not preclude the terns from travelling by geographical landmarks and recognising coasts such as those of Portugal or Morocco.

Experiments with young birds kept in captivity prove that the migratory behaviour and orientation by the sun and stars are instinctive, or preprogrammed in the genes. But the birds do become familiar with the terrain they fly over. Thus very little of this knowledge is acquired, yet it is, nevertheless, information registered in the cells, or matter.

On the other hand, not all of it is inherent, as the following experiment has shown. Eggs of a wild duck living in England were taken to Finland for artificial hatching. When autumn arrived in Finland, the young birds flew south-west towards the Mediterranean over Poland, Czechoslovakia and Austria instead of over France as their parents had done. Genetically, the natural thing for them to have done after the Mediterranean winter would have been to go to England, the territory of their species, but, quite unexpectedly, they took off for their adopted country, Finland, which they reached without difficulty, thereby proving that they had memorised the route.[1]

[1] *Eureka* (annual science digest), Moscow, 1969.

We now make acquaintance with a creature which swims but is not a fish. The playful and intelligent dolphin is a mammal. I remember how these beautiful animals were escorting our liner in the Barrier Reef Sea, always maintaining the same distance and speed as if to say: 'We can swim as fast as you.' Actually they could have travelled twice as fast as the ship. These frolicsome animals have become a serious subject for scientific investigation for it was found that the cell counts in their brains per cubic centimetres are the same as our own. As a matter of fact, their brains are 40 per cent larger than ours, which makes one wonder whether the things we say about the dolphins are as perceptive and far-reaching as what they are probably squeaking about us.

Dolphins do speak, using a language of whistles, clicks, squeaks, squawks and buzzes with high and low frequency components. Their voice boxes can produce twenty identifiable sounds which can be combined into thousands of words.

In one experiment, a dolphin was placed in a tank and taught to push a certain knob with its nose so that a fish would fall into the tank from an automat. The tank was divided by a glass partition into two identical sections including identical knobs. When another dolphin was put into the second section the first dolphin issued a few squawks and the newcomer went straight to the correct knob to get his fish. This was certainly direct communication. It is exactly the same method that we use—we learn from experience and then pass on the knowledge to others.

Dolphins have been taught hundreds of tricks and they enjoy showing off. Scientific teams are now trying new techniques in which the dolphin itself invents new tricks. Some of these impromptu performances are so elaborate that they even surpass those invented by the trainer.

In South Africa there are two dolphins, named Dimple and Haig, who work as beach-guards and life-savers, helping swimmers in difficulties and chasing sharks away.

Dolphins singing at the London Dolphinarium.

Since ancient Greek and Roman times, history abounds in stories of people who have been saved by dolphins, the friends of man. Oppian (second century) wrote: 'It is an offense to the gods to hunt dolphins.'

Unlike man, dolphins do not have to work because their food is all around them. This makes life a continuous game to them—from their intricate underwater ballets in twos and threes to riding the bow wave of passing ships or leaping in and out of the water.

The Soviet expert on dolphins, Dr A. V. Yablokova writes: 'All our attempts at communication may turn out to be futile not because dolphins are devoid of intelligence but because we ourselves have not reached the state of conscious contact with other sentient creatures.'[2]

Dr John C. Lilly, a world authority on the dolphins, admitted that one day he discovered that the dolphins were experimenting with him. 'This gave us all the greatest hope of one day seeing these animals try at least to meet us half way in our efforts at interspecies communication,' he said.[3]

Now what is intelligence? Most scientists and psychologists are in accord that a creature which can anticipate the results of its actions and works towards an aim, exercises reason and is, therefore, an intelligent being. However, there may be many possible ways of storing and processing information, varying according to the different species of intelligent life.

The chimpanzee may not be as clever as the dolphin but he looks human, and anything that resembles man must be intelligent, of course. Would it be possible to teach grammar to a chimpanzee? Professor David Premak of the U.S.A. is convinced that it is. His experimental chimp named Sarah learned to associate different pieces of plastic with words and ideas when they were placed next to objects and specific people in the laboratory. She was clever enough to save herself the trouble of writing the following three sentences in plastic: 'Mary gives Sarah apple,' 'Mary gives Sarah orange,' 'Mary gives Sarah banana.' She abridged them to one sentence: 'Mary gives Sarah apple, orange, banana,' and in this way obtained all the fruit.

The problems of affirmation and negation, yes and no, and of

[2] *Znanie-Sila*, No. 12, 1970 (U.S.S.R.).
[3] R. Stenuit, *The Dolphin—Cousin to Man*, London, 1968.

discrimination, same or different, were also successfully solved by Sarah with her plastic symbols. Finally, the chimp answered the even more difficult question: 'What is red—a colour, shape, size or name of apple?' by replying with the plastic token which

Koko, a Sydney zoo chimpanzee, considers that animals are people.

signified *colour*. This is a major breakthrough in zoopsychology and should make us realise that animals are people, after all.

The chimpanzee is certainly intelligent enough to carry out rational actions. One chimp whose baby tooth was aching, got hold of a pair of pliers and extracted it as a dentist would to the horror of the other animals. Then he stepped on the cause of his pain, while other chimpanzees staged a demonstration of sympathy, angrily kicking the damned tooth around.

One can teach a chimp to use not only a bed and blankets but also that supreme achievement of modern technology—the flushing toilet. A chimp can be even taught to utter words, as Barbara Ford relates in *Science Digest*:[4] 'Gimme drink, please,' with a thumb between lips was a clear indication that the chimp wanted a Coca-Cola. On hearing a dog, he would say: 'Listen, dog.' 'Open flower' meant that he wanted the door opened so that he could go into the garden. But, of course, few chimps speak English, French or German; they have their own language. The philologist George Schwidetzky wrote a book on that subject entitled *Do You speak Chimpanzee?* In the London Zoo he came to the chimpanzee cage and said 'Hello'; there was no reaction of any sort. Then he emitted an undulating 'Oo-oo-oo', and the chimps rushed up to him as if he were a relative or old friend. When two chimps meet each other after separation, they show human-like emotion, holding hands, hugging and kissing each other.

Like humans, chimpanzees are not very interested in working for nothing. They despise those socialistic insects, the bees and ants. The chimps are definitely money conscious, as Dr John Wolfe of the Yale Laboratories of Primate Biology has demonstrated. Vance Packard tells the following story about it in his delightful book *The Human Side of Animals.*

It all began when chimpanzees were taught to use chips in a simple operation, whereby they, imitating the trainer, caught on to the fact that when a white chip was slipped into a special slot machine, out came a grape. But a brass chip was a blank, producing nothing. When the animals realised that only the white chips could be used to obtain grapes, they began hoarding the white chips and throwing away the brass ones.

But this simple Free-Grapes game soon ended and the diabolical mind of the Yale scientist conceived a Work Machine which was constructed, having a large handle with an

[4] *Science Digest*, May, 1970.

8-kilogramme weight attached to it. By exerting themselves and lifting the handle, the apes would get a chip which could then be cashed at the old machine and become a grape. Then yet another variation was introduced which really taxed the chimps' ingenuity. Depending on the amount of work done they received blue, red and yellow chips also. On the chimpanzee money market the blue chip would buy two grapes, the red one produced a drink, and the yellow token entitled the receiver to a piggy-back ride on the psychologist's shoulders. The apes in the laboratory soon became aware that the blue chips had a higher value than the white ones, and that all that was necessary to quench a thirst was a red token. So the chimps started working hard and amassed piles of poker chips. They sat on their fortunes like real misers. However, what this experiment set out to prove, and succeeded in doing was the ability of chimpanzees to differentiate between colour symbols and thus think in almost abstract terms.

There is a most peaceful breed of ape which does not tolerate fights or wars, although his appearance is terrifying, and he is ten times more powerful than man. This is the gorilla. Although capable of squeezing a leopard to death, this giant beast is really quite gentle, as alleged by Dian Fossey, an American scientist. She had first-hand experience of their temperament and habits after having lived among wild gorillas in the heart of Africa for three years.[5]

Unlike most animals, the gorilla tribes often share their territories and even exchange members. The small packs are ruled by benevolent chiefs who direct all activities such as the best times and places for foraging and the most suitable camping or dwelling sites. Gorillas are expert botanists, which is probably due to their diet of fruit, roots and plants. They use at least twenty voice signals to communicate among themselves. It is erroneous to believe that chimpanzees are superior to

[5] *Frontiers*: A Magazine of Natural History, April, 1971.

gorillas in intelligence. Dr Ernst Lang of the Basel Zoo says: 'It's not that gorillas can't do the things that chimps do but just that they are smart enough to do only what they want to do. They are all individualists—having independent and superior personalities.'

The Kansas Recreation and Park Association recently had a painting contest for children. The first prize was won by D. James Orang for his paintings in colour listed as *Train from Tokyo* and *Tornado*. When the painting prize was to be awarded, the judges found out to their amazement and amusement that the artist of these impressionist paintings was orangutan Djakarta Jim from the Topeka Zoo, whose works were entered in the contest, under a nom-de-plume, by the Zoo Director Gary Clarke. 'The paintings seem to come from a creative impulse,' he said.

The aim of these sketches of animal life has been to establish the presence of intelligence among some of our animal relatives which stand reasonably high on the ladder of evolution. All animal evolution for millions of years has been leading to a climax. That climax is man.

7: *The Ascent of Man*

Ever since his appearance on the planetary scene, man has acted as if he meant to replace all the other animals. If the tyrannosaurus had survived, it would have been less dangerous than man. It is the developed, intelligent mind juxtaposed with a lack of consideration for all Nature that makes man such a menace. Yet at the same time his intellect and curiosity have made him a winning card in the game of evolution, thus far.

While man does not descend from the ape, both could have had a common ancestor. His pedigree is lost in the deep roots of evolution but his kinship with the animal kingdom is certain. Man belongs to the vertebrates and his skeleton bears close resemblance to that of fish, frog and dog. All have skulls, spinal columns and fore and hind limbs. The same can be said of the digestive, blood-vascular, respiratory, nervous and reproductive systems as well as the organs of sensation. In the early stages the embryos of man and chick look identical.

How did man appear? Engels gave a convincing reply: 'Man was created by work.' Man is the only creature that can make tools and use them. There are animals and birds that can use tools but none are able to construct them.

It took millions of years for that quadruped which was evolving to become a biped and finally start using stone implements and weapons. Man's greatest weapon was the intelligence by which he rose high above the animals. When the Ice Age arrived he had two possible alternatives open to him—to freeze to death or to fight for survival. So exercising a capacity for choice, he killed beasts for food and used their skins to cover his vulnerability.

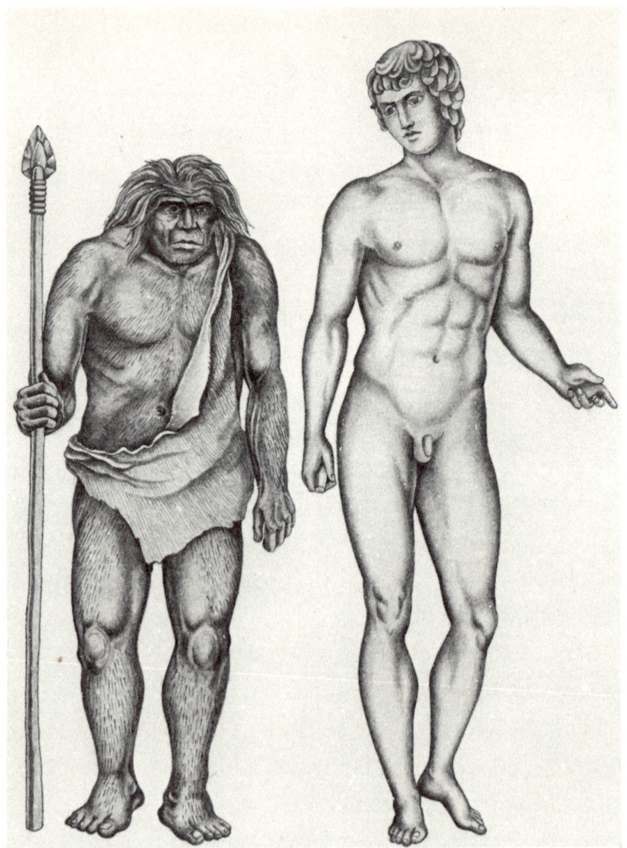

From coarseness to refinement, from ugliness to beauty, from ignorance to intellectuality, so goes the path of evolution.

Birds, insects and animals communicate by means of audible signals. Men perfected this method of communication by sounds and signs, and invented language. The transition stage between the primates and *homo sapiens* must have taken millions of years. The recent discovery of the Leakey skull* poses a possibility that the tool-making man may be millions of years older than is considered at present. The period covered by the

* By Richard Leakey in Kenya in 1972.

history of mankind is completely out of proportion to the period which covers prehistory.

The process of evolution must have been very gradual, occupying long epochs during which man found different uses for his first tools and slowly adapted these to new purposes. The Middle Paleolithic period from about 150,000 B.C. to 35,000 B.C. constituted a turning point. It was probably during this time that man discovered fire and began to use it for heating and cooking.

The first men wielded clubs in their muscular hands to hunt beasts and provide for the tribe. Their restless minds and their spirit of adventure pushed them along the path of ascent. Primitive man began by using his hands only, then learning by trial and error, he acquired the capacity of solving problems in his head, utilising mental pictures of concrete things and abstract ideas. Balzac wrote that: 'At abstraction society begins.' Man knows that he thinks and the human mind can study itself, thereby differentiating him once more from the animal. Man started to watch the sun, moon and stars to notice changes in the weather, and to benefit from these observations. Evolution signifies the growth of reason and the expansion of consciousness, rather than the growth of muscle. The evolutionary adolescence of *homo sapiens* was attained when he began to think in abstract terms.

The hand of the Cosmic Clock moved slowly for thousands and millions of years. An evolutionary programme was already inherent in these ape-like creatures with manes and bright eyes. It was clear that their brains would one day rule the planet and even seek roads to the stars. In the meantime the Galactic Clock ticked on and the stars waited. Aeons passed unnoticed and unsung for Time was then a part of Eternity.

He had to start from scratch and by imagination and mental exercise to invent and discover everything just like a child. For he was a child of the new kingdom of humanity. When at last man conceived of himself with an immortal consciousness and a soul, he discovered himself as if he had looked into a mirror for the first time and had seen his image.

Animals live only in the present but man discovered he could re-live the past and dream about the future. The workings of Nature with her wind, rain and lightning were attributed to acts of the gods, and thus magic and religion came into being. It did not matter in the least whether these myths reflected reality or not. What counted was the new faculty to think abstractly—to generalise, to find the infinite in the finite, the eternal in the temporal.

Vitruvius (first century B.C.) explained the rise of civilisation in these words: 'When men realised that they should walk upright, not with their heads down, and should look upon the magnificence of the world and of the stars,' then culture appeared.

The existence of man on this planet is meaningful only if seen in its evolutionary perspective. He has emerged from the primates and is climbing towards a superior stage of life and consciousness. The trend of man's development points towards the abstract. When early man created mythology, religion, mathematics, astronomy, music and art, he became true man. It should be kept in mind that the abstract does not necessarily mean the unreal. Einstein's formulae are not easy to understand-in their abstractness, yet they form the basis of the atomic bomb which is awful in its reality.

The *Pyramid Texts* of Egypt contain these beautiful expressions of man's inner understanding of the intangible principles of the universe:

'My body to the earth, my soul to the sky' . . . In the *Book of the Dead* we find this passage: 'Mine is yesterday, I know tomorrow . . .' In the Pharaoh Pepi's funerary inscription which is more than 4,500 years old, there are these inspiring words: 'To the sky . . . To the sky . . . To the great seat among the gods!' And the Hymn to the Sun of Amenhotep contains both poesy and scientific truth: 'Though thou art far away, thy rays are upon the earth, thy love is great and mighty.'

Civilisation commenced when the small bands of men and women, who roamed the wilds, joined forces and settled down in larger groups forming villages; there, instead of gathering grain and roots, man began to plant and grow them; instead of hunting animals, he decided to domesticate them. After that came barter and trade and with it the building of towns and cities.

History and archaeology draw a colourful picture of the early civilisations of Mesopotamia, the Nile Valley, Greece, Rome, China, India, Mexico and Peru. There have been dark epochs of brutal wars as well as brilliant eras of culture. Man searched for truth in the invisible world and created religion. He sought it in the visible world and discovered art and science. His horizons expanded greatly and psychologically he was as different from his prehistoric ancestors as the hairy, stocky Neanderthal man was different from a marble statue of handsome Apollo.

The atoms of our bodies come from the volcanoes, seas, air and rocks of this planet, lightning, the sun and the stars. This may be the reason man has an intuitive feeling of infinity and eternity for he is made up of eternal atoms. Let us quote Teilhard de Chardin: 'I give the name of cosmic sense to the more or less confused affinity that binds us psychologically to the All which envelops us.'[1]

Admittedly, a great deal more could be said about man. However, this work is written from an extraterrestrial position, from which man does not appear to be very conspicuous. Moreover, on a time scale which deals with hundreds of millions of years the existence of the human race on this planet has been extremely brief.

[1] T. de Chardin, *Let Me Explain*, London, 1970.

Man is only one letter in an evolutionary sentence that has no beginning or end. Picus de Mirandola defined man as 'the intermediary between creatures, familiar with the gods above as he is lord of the beings below'.

The terrestrial breeding ground of evolution has already produced men approaching a supra-human stage. Plato was able to raise himself to an archetypal world of pure ideas, Buddha received a revelation of the timelessness of the universe, St Francis of Assisi felt his unity with Nature to such an extent that he spoke to 'Brother Sun and Sister Moon', Leonardo da Vinci made mental pictures live, Beethoven thought with music, Einstein discovered a new dimension in the old universe. These are just a few examples of men who were centuries ahead of their contemporaries, each in his own way.

If this view of man as a step on the ladder of infinite progression is rejected, then the other alternative would be to regard him as the final product of evolution in the universe. Until this illusion of his grandeur is dissipated, man will go on beating his chest and shouting: 'I am the best and the most intelligent being in all creation.' Do these bombastic words bring laughter in the stars?

The story of man could be long and glorious, or short and infamous. The first men received the fire of Prometheus. The last men have opened the box of Pandora and released the fire of the atom which could destroy this planet. Will mankind commit a global suicide, or plant on Earth a new garden of Eden? The choice is his.

8: *Beyond Mankind*

Wherever there are organs of perception, however elementary, there must be a perceiving consciousness, as without it the organs would be useless. The mystery of the universe is the self-perfection of matter that ultimately cognises itself by a property which may be inherent in it, or consciousness, mind, intelligence.

'Had Nature created sentiency only for the purpose of ensuring a food supply for living forms, she might be considered excessive in her abundance,' writes the Soviet scientist I. M. Zabelin.[1] While not denying the fact of the importance of 'self-feed' to rational beings, he envisages a more sublime goal for mankind.

In a flight of imagination, we might scan the future and see man creating a just planetary society without wars and misery. Taking the optimistic viewpoint, one could imagine that man on this planet might achieve an utopia some time in the future. But now our question is—can man become something greater than himself? As the primate gradually evolved into modern man, which is an entirely new species, so must man by mutation develop into a superior being.

Physiologically, man is changing. Global statistics show that the teenagers of today are a few centimetres taller than were their counterparts in the second half of the nineteenth century. Children attain adolescence one or two years earlier than their forebears. Intermarriage between various migrating ethnical groups has created mutations. Social factors and a faster tempo

[1] I. Adabashev, *Mirovye Zagadki Segodnia* (Riddles of the Universe Today), Moscow, 1969.

of life are adversely influencing the nervous system of man, while at the same time stimulating his mental altertness to an unparalleled degree.

The mere fact of man's evolution from a stocky, clumsy humanoid with a low forehead to modern man, compels one to believe that in a few million years, or perhaps even sooner, the human being should be as much superior to present-day man as he is to the man-ape.

Nietsche said: 'Man is something that shall be surpassed.' Without a realisation that man has sprung from the lower forms of life and is heading for higher ones, his existence on Earth would seem to be devoid of significance. To believe that man is the crown of evolution and that Nature has laboured for aeons to produce present humanity as the apex of her achievements, would appear to take all the rationale out of evolution.

Man is only a bridge, one end of which is fixed to the cell and the other lost in the stars. With this vision of a superhuman goal for man, his true mission will come into focus. The laboratory of human civilisation will eventually produce the next evolutionary specimen—the ultra-man.

How can superman be defined? If concrete mind has been the weapon of man, then abstract mind and intuition will be the tool of superman. This new offspring of Nature is not going to be a marching warrior but 'an astronaut of spirit' with wide cosmic horizons, who in the words of Blake 'will hold infinity in the palm of his hand and eternity in an hour'. In the long process of the distillation of consciousness, man should reach a point where he can sense his affinity with the cosmos. By this transmutation of the concrete mind into the abstract, the earth-man becomes star-man.

The concept of a being more elevated than man is fully in accord with the most broadminded scientific opinion, to which dead-ends in evolution being illogical, are simply inconceivable.

Neither is it foreign to religion, which has always believed in superior beings, whether considered angels or gods.

No one has depicted the infinite progression of life more poetically than the Sufi thinker Jalalu'ddin Rumi (1207–1273):

I died as a mineral and became a plant,
I died as plant and rose to animal,
I died as animal and I was man.
Why should I fear? When I was less by dieing?
Yet once more shall I die as man
To soar with angels blest but even from angelhood
I must pass on . . .

To become superman, to reach another evolutionary stage, man must ascend to a higher level of both consciousness and physical organisation. The trend of mankind is towards an ultra-human sphere of cosmic awareness. If this presumption is denied, then man would have to be considered a static entity already overflowing with wisdom and shining in beauty, with no need of perfecting.

Humanity has now reached crossroads where its high intelligence can be misused since, instead of turning it towards constructive purposes, it has been developed into a furor for destruction. Let us hope that this situation is transitory, and that it may indicate nothing more than a painful stage in our growth.

A time will come when mankind will have to decide whether or not to co-operate with Nature in her strenuous efforts to turn us into supermen. Making allowance for accelerations which are not uncommon in evolution, the transition into the supra-human may take tens and even hundreds of thousands of years, assuming, of course, that in the meantime we do not choose to return to the level of the primates.

The normal process of evolution together with conscious effort may have already created races of ultra-men in other solar systems. The great German Darwinist Ernst Haeckel speculated brilliantly upon this subject: 'It is possible that other planets have produced other types of higher plants and animals which are unknown on our Earth. Perhaps from some higher animal stem which is superior to the vertebrate in formation, higher beings have arisen who far transcend us earthly men in intelligence.'

Perhaps cosmic life and mind could be compared to a ladder in the form of a spiral that eternally progresses. It could have curves of temporary regression without endings, as there are no fullstops in infinity. And where lies the destiny of future superman? Only in space, for he must leave this planet in mind and body to become a truly cosmic being. Although earthmen have made trips to the moon, there exists a wide disparity between that technology geared to space exploration and the mass psychology of populations. One is Copernican, the other still Ptolemean or geocentric. Superman will be a reality only when man becomes aware of his cosmic origin and affinity with Nature.

The ancient Chinese and Egyptians already had an inkling of this, as about 2,000 years ago the Chinese philosopher Wang Ch'hung, in his book *Lun Heng*, wrote that: 'Man lives on the Earth's surface like lice in the folds of a garment.' And on the Pharaoh Menkaure's tomb there is the following inscription: 'Born of the sky, conceived of Nut (heaven), heir of Geb (earth).' Compared to the size of our planet man is certainly small but his mind carries the seed of a new universal being. The day may come when man will lay new roads in space, taking him out of this solar system. He will also explore the depths of inner space, his mind, and in the course of centuries or millennia become a galaxy of spirit. Then, and only then, will superman be born on this, the third planet of this solar system.

Few can grasp the fact that the impact of the present Space

Age on humanity is going to be greater than the Christian revolution of the Roman Empire, the Holy Wars of Islam, the Crusades or the Renaissance. The egress of man into space should raise his morality and level of consciousness so that in the long run mankind as a whole will gain spiritually from this material achievement. This conviction can be substantiated by the psychological reactions of American astronauts in space.

'I am not the same man. None of us are. I felt part of everyone and everything sweeping past me below,' said Apollo 9 Astronaut Rusty Schweickart.

'Something happens to you out there—you develop an instant global consciousness, a people orientation, an intense dissatisfaction with the state of the world and a compulsion to do something about it,' remarked Apollo 14 Astronaut Ed Mitchell.

Apollo 15 Astronaut Jim Irwin 'was deeply moved by the beauty of the lunar mountains and felt the presence of God'.

'I was overwhelmed by the certainty that what I was witnessing was part of the universality of God,' stated Apollo 16 Astronaut Charles Duke.

'I didn't feel like a giant. I felt very, very small,' said Apollo 11 Astronaut Neil Armstrong.

Apollo 8 Astronaut Bill Anders experienced 'feelings about humanity and human needs that I never had before'.

'You don't look down at the world as an American but as a human being,' remarked Apollo 10 Astronaut Tom Stafford.[2]

Michael Collins, another American astronaut, wrote that 'So I have a personal, simple message to pass on: There is only one Earth. It is a tiny, precious stone. Let us treasure it; for there is not another one.'[3]

These are the very emotions and intellectual attitudes that will make supermen out of earthmen.

[2] TIME, 11 December, 1972.

[3] *Guideposts* (magazine), New York, 1971.

Throughout history many illumined minds have had glimpses of this cosmic consciousness which will characterise the coming superman. Pythagoras wrote this verse on his intuitive perception of that consciousness:

> I was lifted high into the heavens, and there became a star,
> I sang with all my starry sisters, and we danced around the throne of Time.

And concerning this same expansion of consciousness which the man of tomorrow is bound to experience when confronted with life in infinite space, the Chinese philosopher Chuang Tzu expressed this thought over 2,500 years ago:

> Now that you have emerged from your narrow sphere,
> And have seen the great ocean,
> You know your own significance.
> I can speak to you of great principles—
> Dimensions are limitless, Time is endless.

One of the basic changes in *homo sapiens* when he becomes the superior planetary dweller of the future, will be a different sense of time. In this Einsteinian century time has become relative, and dependent upon the observer. The angle of temporal vision of coming supermankind might be immensely wider than our own.

One thing is certain—after exhausting all the potentialities of the concrete mind, man will ultimately turn to abstract thought, which will lead to a richer mental life. In the initial stages there may be only a few hundred of these pioneers of a new species but later thousands and then millions will join them. Let us not forget that it has all happened before when man first emerged from his beginnings! *Man into Superman* will be the next act in the drama of terrestrial evolution.

The noted British scholar Dr W. Y. Evans Wentz puts forward the following question: 'Are there members of the human race who have reached, as Milarepa is believed to have done, the height of such spiritual and physical evolution as this planet admits, and who, being as it were, a species apart from other human beings, are possessed of mastery over natural forces as yet undiscovered, but probably suspected, by Western Science?'[4]

It would be possible to imagine an older galactic civilisation in which the average level of consciousness would gravitate towards the heights reached on earth by Christ, Buddha, Pythagoras, Plato or Einstein. A planet of geniuses is certainly an evolutionary possibility.

[4] W. Y. Evans Wentz, *Tibet's Great Yogi Milarepa*, Oxford, 1969.

9: *The Stairway to Perfection*

Cosmic evolution is a fundamental factor by which everything is affected—atoms, cells, biological organisms, planets and stars. A study of paradigms in the physical and biological fields shows that smaller units build larger agglomerations in a strict hierarchial order. Moreover, the number of steps from an atom to a galaxy is limited. A molecule is a unit made up of atoms, a cell is composed of molecules, then these cells amalgamate into a larger unit—an organism, and organisms form a particular species. Next is a planet such as the Earth, a member of a solar system, which in its turn is only a particle of the Milky Way galaxy.

It is said that man is halfway in physical size between an atom and a star. In the world of the infinitely great, our mental gaze can not penetrate beyond the meta-galaxy; in the atomic world it stops before the proton and its particles. The atom is similar to a miniature planetary system in which the nucleus plays the role of the sun and the electrons replace the planets.

A century ago most scholars saw nothing but inert matter under their feet and solid or gaseous astronomical bodies over their heads. To them the universe was largely dead. With the discovery of the throbbing atom, the heart of all matter, and of stellar evolution in space where everything is moving at fantastic speeds, the picture of a static world was torn to shreds.

One of the qualifications of life is movement. Do either atomic particles or stars lack this? No true scientist can now say that the universe, with its myriads of whirling atoms and galaxies, is lifeless. It is currently accepted that biological evolution is but a part of cosmic evolution. All the chemical

Superhuman life systems must exist in the universe if man is not the last word in cosmic evolution. (By William Blake, 18th century poet and artist.)

elements have evolved and are still evolving from the simplest and lightest of atoms—hydrogen and helium. In the astronomical universe the urge towards complexification exists as much as it does in biological forms. Diffuse clouds of hydrogen first condense into nebulae, then into stars, star clusters, galaxies and meta-galaxies. When the stars exhaust their nuclear fuel and cease to radiate, they die. Stars and worlds are born and die like living things on this planet. But it is evident that the universe of universes must have regenerative and regulating system in order that life may go on without end.

Since the law of complexification in biological evolution eventually produces consciousness, can we say with certainty that this law does not operate in cosmic evolution? Can we be sure that above man there is a void as far as sentience is concerned? This anthropocentrism confining intelligence to a narrow band in the ladder of evolution must be questioned. Does the stairway of evolution cease abruptly, and are we, human beings, standing on its top step? And suppose there are further and higher steps?

We are now in the land of Brobdingnag of Gulliver's Travels—the domain of the Titans. In the world of Lilliput we were giants ourselves as we studied the life of the ants and bees. But we shall find this new unexplored realm to be quite awesome, as the beings who inhabit it are colossal, and we are but planetary microbes. We must temporarily leave our astronomical provincialism as this is not a land for intellectual dwarfs.

From the dawn of civilisation man has believed that the sun, moon and stars were the physical vehicles of sentient beings.

'The gods sit on their stellar thrones,' said Lao-Tzu, the founder of Taoism. Under the name of *Celestial Honoured Ones* Taoists worshipped the spirits of the stars. Ancient Chinese writings state that 'Heaven is the dwelling of sidereal divinities and each god has his own palace'. The Aztec priests recorded their belief that where ordinary men 'behold a sun, we perceive a deity'. Pythagoras considered all astronomical bodies as living and conscious beings and said that their minds were so magnificent they were worthy of man's adoration and imitation. The cult of the sun, moon and stars was continued in Egypt for 5,000 years.

The stellar religion was universal in antiquity. 'We tend to think of the stars as mere bodies or items arranged in order quite without soul or life; we ought rather to regard them as possessed of life and activity,' wrote Aristotle.

Babylon had its cult of Shamash, Sin, Nergal, Nebo, Marduk, Ishtar and Ninib. In India the astronomical Pantheon is comprised of Surya, Chandra, Karktikeya, Budha, Brahaspati, Sukra and Sani. In ancient Greece the divinities were known as Helios, Selene, Ares, Hermes, Zeus, Aphrodite and Cronos. In Rome they bore the names of Apollo, Luna, Mars, Mercury, Jupiter, Venus and Saturn. The planetary gods who were worshipped by these early civilisations all identifying themselves with the sun, moon, Mars, Mercury, Jupiter, Venus and Saturn, have been handed down to us in the names of the days of our week—Sunday, Monday, Tuesday, Wednesday, Thursday, Friday and Saturday.*

Cicero attributed divinity and intelligence to the stars and planets, an opinion that was shared by Rome and Greece alike. The early Fathers of the Church such as St Augustine and Origen agreed that every star was under the guidance of an angel. St Thomas Aquinas wrote: 'I do not remember having ever met in the works of saints or philosophers a denial that the planets are guided by spiritual beings.' Up to the eighth century this belief remained incorporated in Christian dogma. Even as late as the sixteenth century revivals of this ancient tenet occurred in the Catholic Church. A papal bull issued by Pius V, sanctioning in Spain the veneration of the Seven Spirits of God, declared: 'One could never exalt too much these seven Rectors of the world, figured by the Seven Planets. It looks consoling and augurs well for this century that by the grace of God, the cult of these seven ardent lights, and these seven stars, was regaining all its lustre in the Christian Republic.'

The Hebrew Kabala embodies the doctrine of a celestial hierarchy, beginning with the planetary angels—Michael, Gabriel, Samael, Raphael, Zadkiel, Anael and Cassiel. To this cosmic hierarchy are added the Seraphim, Cherubim, Thrones, Dominions, Virtues, Powers, Principalities, Archangels and

* Dimanche, Lundi, Mardi, Mercredi, Jeudi, Vendredi, Samedi.

Angels, embracing a wide range of supra-human intelligences. This secret teaching claims to possess the *heavenly alphabet*, or the science of the correspondences and correlations of these beings not only with the constellations but with every type and kind of spiritual, mental and physical phenomena at all levels throughout the macrocosmic and microcosmic universe.

The ancient Brahmins compared this planet with a man—the forests were likened to the hair of man, the seas to his sweat and the volcanic lava to his blood.

Everything that has been said so far may belong to mythology rather than to science. On the other hand, it is undeniable that the biosphere of this planet consists of living matter. The seas and soil are swarming with life and so is the air. However, is the life layer that envelops the Earth only skin deep?

Dr Gustav Fechner, a German scholar of the nineteenth century, thought that the entire Earth is alive and has its own collective consciousness. To cite his own words:

> The view that in the heavenly bodies there dwell heavenly souls is so strange to the world of today that we can not but wonder why it could seem so natural to an earlier world. Man is the measure of the world. Just as man is the starting point and the point of reference for belief in the animate character of all other creatures, so is the animated Earth the starting point and the point of reference for belief in the animate character of all other stars which inhabit with it the same heaven.[1]

Another scientist echoed this assumption around the same time. This was Dr Thomas Huxley, whose struggle for Darwinism is recorded in the history of science: 'For anything that can be proved to the contrary, there may be, somewhere, a finite being or beings, who can play with the solar system as a

[1] G. Fechner, *Religion of a Scientist*, New York, 1946.

child plays with a toy.'[2] On the background of Huxley's violent clash with Samuel Wilberforce, Bishop of Oxford, in 1860 no one can accuse him of excessive religiousity, yet as an evolutionist he was capable of imagining superior entities on the stairway to perfection.

Herein lies the great paradox—although the ancient Greeks were right in saying that the gods were created by fear and the priesthood sustained by superstition, yet one must be careful not to write off all manifestation of religion. Quite possibly, atheism may be as great a fallacy as religion. Truth is found at the centre of the Golden Mean.

Perhaps man is as yet too undeveloped to comprehend this field of supra-human evolution. To date he possesses no instruments to test this hypothesis, aside from his mental faculties.

About 300 years ago Leibniz proposed his Monadic Theory, according to which 'all complex formations are composed of simpler force units'. This is true at atomic or cellular level and it may be fact on higher planes as well. Leibniz proposed the existence of aggregates of monads, the cohesion of which was maintained by the predominating unit—a centre that held the organism or the conglomerated mass in balance. Actually Leibniz's theory is supported by a law of physics and biology according to which 'the whole is greater than the parts of which it is composed'. Under natural circumstances a living cell can not live if removed from the body. Likewise a portion of stellar substance separated from the star would evaporate into space and grow cold. This results from the interaction of a particle with the total mass. The generation of energy inside a star depends upon the mass of substance outside its centre. And the same is true of an organism.[3]

[2] C. Bibby, *The Essence of T. H. Huxley*, London, 1967.

[3] *Illustrated London News*, 25 February, 1961 (Dr M. W. Ovenden—Is there Life on the Planets?).

The human body is composed of organs and cells. When a virus or germ attacks it, the white blood cells rush to repel the enemy as if they were sentient entities. When an organ is transplanted, certain cells sense the outsider and pounce upon it, a challenge that has confronted medical science and that has been dealt with successfully several times recently, despite the failures encountered in the interim. When it is possible to fool these blood soldiers completely, surgeons will be able to transplant anything from anybody in the same blood group. All these microscopic soldier cells are concerned with is the integrity of the organism, that is the purpose for which they are programmed.

When Lowenhoek first looked at a drop of water through his newly invented microscope, he was stunned to see a whole new world consisting of microbes. Some 50 years ago when astronomers at Mount Palomar examined the photographs of nebulae, they realised that they had discovered hitherto unknown island universes of stars. The misty nebulae were resolved into rings, spirals and ovals reminiscent of Lowenhoek's germs.

The concept of greater aggregates composed of lesser ones allows one to perceive the whole Milky Way galaxy as an integral assemblage, not only on the astronomical level but on the intelligence level as well. Perhaps a good simile would be that of a swarm of bees, with the 100,000 million stars of our galaxy playing the role of the bees. As a beehive has one mind, one purpose and one life, so may the Milky Way galaxy also be one system. It is practically impossible to envision the scope of a cosmic intelligence of such grandeur.

Fred Hoyle's science fiction novel *Black Cloud* tells the story of an astronomical formation containing an intelligence. When scientists established contact with the 'cloud', it made the following remark: 'Probably you have wondered whether a larger-scale intelligence than your own exists. Now you know that it does. In like fashion I ponder on the existence of a larger-scale intelligence than myself.' A similar thought has been expressed by the noted British astronomer V. A. Firsoff in his book *Life, Mind and Galaxies:* 'A comparatively complicated structure such as a galaxy which bears a resemblance to an organism and has a kind of metabolism, nuclear instead of chemical, must be expected to have some kind of mind, perhaps of a high order.'

Professor J. B. S. Haldane, foremost British biologist, theorises that there might be creatures in the universe as vastly more intelligent than man as man is more intelligent than the earthworm. A being on such an exalted plane may be 'managing the affairs of our world, or perhaps our solar system or even our galaxy, while only revealing its existence directly to a select few human beings'.[4] Haldane proposes a project to collect data which would give some evidence as to the development of our galaxy, in order to determine whether it shows any intelligent characteristics. It only remains to ask whether man's limited and comparatively low level of intelligence would be capable of recognising any cosmic intelligent characteristics? It may well be that astronomers are observing these phenomena without understanding their true meaning.

From the foregoing it is clear than an hypothesis which would consider a galaxy to be not only an astronomical formation but also an organism, is very much within the framework of legitimate scientific thought. 'We have no inkling of our position in the hierarchy of the universe. There may be intellects among the stars as vast as worlds, or suns,' wrote Arthur C. Clarke in *Profiles of the Future.*

Imaginatively, Teilhard de Chardin spoke of *thinking stellar units.* Consequently, somewhere in the cosmos there may be intelligences who view us with the same air of justified superiority as doctors examine bacteria under a microscope. And

[4] J. B. S. Haldane, *Science and Life*, London, 1968.

one may also wonder whether there exists some sort of ecological cycle within the entire cosmic system.

Nature insists on excellence and perfection throughout the whole course of evolution. She builds greater units out of smaller ones. During the longer cycles, which to us seem like eternity, the lower forms of life and consciousness are raised to a higher level of cosmic awareness. Is the purpose of the material universe to see its own beauty and listen to its own symphony?

The view of Leibniz of a hierarchy of monads is far from absurd. Sentient humans may have the same relation to cosmic pantheons as cells to the body of man. It is more than likely that communication between conscious cosmic units would be maintained on their own level or frequency. By stretching our imagination a little bit further we could conceive of these beings as functioning simultaneously in a parallel universe. This is not science fiction. Professor G. I. Naan of the Estonian Academy of Sciences (U.S.S.R.) believes that our world exists parallel to an invisible world of a different polarity. In that universe all polarities are reversed and a particle with a positive charge here would carry a negative one there.

A neutrino* mind would observe our astronomical universe only as thin patches of mist. To us it would be an impossible world where a particle does not obey any gravitational or electromagnetic forces. It has no mass, a different time and can reverse its time arrow.

Could there be a mental plane of this type where nothing exists other than thought vibrations and forms? The power of thought has always been underestimated. Yet everything around us—cities, buildings, cars, bridges, planes, works of art, music—was first created by the mind.

Teilhard de Chardin believed that there is 'a continuous

* *Neutrino*—an uncharged elementary particle having a mass less than $\frac{1}{10}$ that of the electron.

layer of thought round the Earth—the Noosphere'. This would seem impossible at first, but let us recollect that the whole world of microbes was entirely unknown before the invention of the microscope. Radio waves and cosmic rays had been bombarding the Earth for aeons before we discovered them.

By analogy let us suppose there is an extensive force-field around our planet on a non-electromagnetic mental frequency. Telepathy, a non-electromagnetic phenomenon, would then find a scientific explanation. This Noosphere may be as real as the Heaviside-Kennelly or Appleton layers that reflect radio waves. Columbia University's Gerald Feinberg suggests that psychic or mental messages may be transmitted on the level of as yet undiscovered elementary particles which he calls 'psychons' or 'mindons'.

Can thought affect matter? Professor Rémy Chauvin of Strasbourg University, using an uranium isotope, a Geiger counter and several assistants, performed a fantastic experiment proving that mind can indeed influence matter. Professor Chauvin asked the experimenters to focus their thoughts during the first minute on accelerating radioactive disintegration, to concentrate on slowing it down by willpower during the second minute and to turn off their thoughts for the third minute. The Geiger counter automatically stopped every minute, so when the results were checked, the participants of the test could hardly believe their eyes—they had actually succeeded in controlling isotopic disintegration by their concerted mental power.[5]

If Teilhard de Chardin's Noosphere, or the Akasa of the Brahmin philosophy does exist, then communication between the members of stellar Olympuses may be conducted on the Noosphere plane which would be on a subatomic level.

We must now come down to earth as the ascent to the lofty heights of Mount Imagination can be dangerous if not controlled by logic.

[5] *Znanie-Sila* magazine, No. 9, 1967 (U.S.S.R.).

10: *To Err or Not to Err?*

A fundamental truth must be observed with great clarity—all human knowledge has a relative value because it changes and expands with time. This theory of the Relativity of Knowledge is a logical conclusion, emerging from the relativist concept of Einstein. It is plain relativist physics to say that in the sea of infinity which surrounds us, there is no centre. It is an axiom that the ocean of mankind has no centre either. There are no chosen peoples or any one man possessing all wisdom and authority.

The advancement of human thought has ever been dependent upon those who err less and know more. World history abounds in doctrines containing little knowledge and many errors. These have thrived on ignorance enabling the forming of establishments. To expose the fallacies of these systems has never been a safe or pleasant business.

Truth is relative. It is essentially maximum knowledge properly expressed. When a belief is challenged, sooner or later an avalanche of information discrediting it can sweep it away. The story of human error is a long tragi-comedy which is not yet ended.

On 22 June, 1633 the Inquisitors-General of the Catholic Church sentenced Galileo Galilei to life imprisonment for holding the Copernican opinion that the sun is stationary and the Earth revolves around it. The condemnation stated that his belief 'is absurd and false philosophically and formally heretical because it is expressly contrary to Holy Scripture'. In this particular instance the Lutherans agreed with the Catholic

Church because of Martin Luther's statement written a century earlier that: 'The world with all its creatures was created in six days and the biblical account speaks neither allegorically nor figuratively.'

Of the ecclesiastical tribunal which consisted of seven cardinals, three refused to sign the verdict, all to their credit. In actual fact, the next day the sentence was commuted to house arrest by order of the Pope.

Can the cardinals be freed from all blame because of a cosmological error in the Bible? The first page of Genesis alludes to the seven days of creation. For the sake of simplicity, let us call them Monday, Tuesday, Wednesday and so on. On Wednesday dry land was separated from the seas and the earth began to bring forth grass. On Thursday, 'to give light upon the Earth', God created the sun and the moon as well as the stars. Therefore, in this sequence of creation the Earth 'with the grass' had existed one whole day before the appearance of the sun, moon and stars. No wonder the cardinals had to condemn poor old Galileo for professing the pagan heliocentric theory of ancient Greek philosophers such as Anaximander or Aristarchus; it did not fit into the geocentric cosmogony of Moses which placed the Earth first in the order of creation.

The Church could have saved itself much embarrassment had it relied more on its tradition and the writings of its own scholars such as Cardinal Nicholas de Cusa who long before Copernicus and Galileo anticipated modern scientific views. In his book *De Docta Ignorantia*, or *Of Learned Ignorance*, published in 1440, Cardinal de Cusa wrote:

> This world is like a vast machine, having its centre everywhere, and its circumference nowhere. Hence, this Earth not being in the centre, can not therefore be motionless, and though it is far smaller than the sun, one must not conclude for all that, that she is worse. One can not see whether its

inhabitants are superior to those who dwell nearer to the sun, or in other stars, as sidereal space can not be deprived of inhabitants.

The religious establishment is not by any means the only orthodoxy to have made monumental errors in the past. In the late 1890s a foremost British scientist, Lord Kelvin, declined an invitation to be a member of the Royal Aeronautical Society on the grounds that heavier-than-air machines would never fly. The American astronomer Simon Newcomb said in 1903 that winged aircraft were an utter impossibility and submitted mathematical formulae to prove his point. The invention of the aeroplane by the Wright Brothers successfully contested his equations. This did not prevent another astronomer William Pickering from expressing his scepticism, a few years later, about aeroplanes ever crossing the Atlantic, by stating: 'It seems safe to say that such ideas must be wholly visionary.' In 1920 the *New York Times* ridiculed Robert Goddard, the American rocket pioneer, for insisting that rockets could travel in space. 'He seems to lack the knowledge about gravity ladled out daily in our high schools,' said the editorial.

When a strong political establishment, based on a fixed ideology as rigid as the most bigoted religion, discovers that a progressive scientist is defying its dogma, it removes him. This is what happened to the Soviet geneticist Nicholas Vavilov, member of the Royal Society,* who opposed Lysenko in 1940. Lysenko did not recognise genetics and put too much emphasis on environment, expressing the Lamarckian version of evolution. He claimed that wheat grown in suitable conditions could become rye, which is equivalent to saying that dogs living in the wilds could give birth to foxes without breeding with them. Stalin's loyal party members supported Lysenko, with the result that a scientific controversy became a tragedy—

* The oldest scientific body in Great Britain.

Professor Nicholas Vavilov died in a concentration camp in Magodan, Siberia in 1943.* Russian biology was set back twenty years as a result of this interference of politicians with science.

And now to Relativity. When the Theory was first brought to public attention about 50 years ago, one official Russian publication contained an article by A. Maximov, stating that Soviet science was against it for the simple reason that 'the Relativity Principle served exclusively religious and metaphysical tendencies'. In 1926 the Archbishop of Boston wrote these lines about the Theory of Relativity: 'I tell you that this theory became outmoded because it was mainly materialistic and therefore unable to stand the test of time.' Is it not ridiculous that in one case Einstein is accused of spreading religion and metaphysics and in another materialism?

The cornerstone idea of all established systems—'We can't be wrong,' has been shattered too many times. It is the common practice of all dogmatic establishments, whether they are political, scientific or religious, to play down a discovery as long as they can. As Rousseau once said: 'Children are afraid of darkness, while we are often frightened by light.'

After the construction of his telescope in Florence, Galileo invited the astronomers of the city to witness a demonstration. In his letter to Kepler he described what had happened: 'My dear Kepler, what would you say of the learned here who, replete with the pertinacity of the asp, have steadfastly refused to cast a glance through the telescope? Shall we laugh or shall we cry?'

If clashes of ideas were confined merely to the forum, these debates would be most amusing. Unfortunately, human beings have been tortured and slain for opinions. In Spain, the Grand Inquisitor Torquemada burned 10,000 alleged heretics and

* He is not to be confused with Professor S. A. Vavilov, who prospered in the Stalinist era.

tortured 100,000 more. The victims of fanaticism in the 20th century have been frighteningly more numerous than 500 years ago.

Many fallacies are due to *sensorialism*, or the dependence on the impressions of the senses, without making the necessary corrections. The attitude 'I believe only what I can see and touch', existed among the ignorant as much as among the learned. People thought the sun moved around the Earth because it seemed to do so. The Earth appeared to be flat and thus man thought it was. But in our day we have been taught that the Earth is a sphere that moves around the sun, so now we make suitable corrections to the information which comes from the sense organs.

Scepticism is quite often less an expression of intelligence than of prejudice. It is an attitude or an opinion and, therefore, invalid when faced with facts opposing it. Nevertheless, it is essential to maintain a critical yet open-minded stand at all times. The sensorialists of 100 or 200 years ago rejected the atomic structure of matter because no one had seen an atom. That kind of scepticism tends to breed ignorance.

In the positivist era 100 years ago common sense was considered to be a criterium of human wisdom. Gustav Naan, a physicist of Estonia (U.S.S.R.), made the following observation on the subject of sensorialism: 'What is common sense? It is the epitome of our experience and superstition. It is the sum total, on the one hand, of what has been verified by experience, and on the other, of ignorance regarded as knowledge.'

Some materialists believed in nothing that was not as tangible as a rock and were certain that an atom was something comparable to a grain of sand—solid and concrete. But how solid is solid, and how real is real? The sensorialist criteria of the nineteenth-century mechanistic materialism were disposed of by Einstein and Rutherford, and are being buried deeper by nuclear physics every year. We now know that the so-called solid atom, the building brick of all matter, is 99·999 per cent empty space! If electrons could fall on to the nucleus and thus obliterate all of the space between them, the Earth would shrink to the size of a football!

Recent discoveries in physics have shown that some atomic particles, such as the positron, can reverse their time direction and move from the future into the past. Thus matter is mostly a vacuum, electrons sometimes a wave and sometimes a particle, and positrons are born tomorrow to live yesterday. Are these the 'concrete realities' of positivist, mechanistic materialism?

As man's mind is finite and the universe infinite, all human knowledge, therefore, is only partial and relative, and no school of thought can have a monopoly on truth. The old Chinese adage 'Let one hundred schools contend' still holds good today. We know that authority is not a test of truth as it has been invalidated many times. In quest of good judgment and an intellectual technique which would help us to avoid fallacies, let us travel into the distant past to the Golden Age of Greece around the middle of the first millennium before our era, when philosophy meant scientific speculation.

Early in the sixth century B.C. an aged Greek named Thales boarded a Phoenician ship at Miletus on his way to Egypt where he planned to study in the temples of Thoth or Hermes, the god of science. Upon his return he astonished the citizens of his home town by his declaration that 'the sun and stars are merely balls of fire of colossal size'.

His pupil Anaximander of Miletus believed that the world began as an undifferentiated mass from primordial substance which had always existed. He even described how this matter gradually condensed from gas into liquid and solid states, and spoke of the effects of temperature on the creation of the world.

Heraclitus left wealth in order to gain wisdom at the temple of Ephesus. 'All things forever flow and change, except the law of change,' he later taught. 'Contradiction and struggle is the cause of progress and growth,' was one of his basic arguments. 'The universe is periodically flaring up and dying down. It has been created by no one, and it has had no beginning nor will it ever have an end,' is another of Heraclitus's profound paradoxical thoughts which anticipated modern scientific views. The law of the conservation of energy in physics is the clearest hint that the infinite universe is immortal.

The whole cosmos is built of paradoxes. The visible world is made up of invisible atoms. Mind comes out of matter. Time is within eternity. Man is the one and the many—a personality and a conglomeration of billions of cells.

Many years after Heraclitus another philosopher pronounced these words: 'Fools, how can anything come from nothing? Motion and stability, life and death are merely two faces of one reality.' Thus spoke Empedocles. It was only in the eighteenth century that Leibniz and Lavoisier proved this law of the conservation of matter. Atomics has estimated that the life-spans of certain atomic particles are extremely brief but that each electron, proton and neutrino has always existed. The universe was created out of these building blocks and certainly not out of nothing.

Empedocles, known as 'the gloomy philosopher', was followed by 'the laughing philosopher'—Democritus of Abdera. He was a genius who was capable of thinking in the same terms as the scientists of the distant future for it was he who said 'in reality there is nothing but atoms and space'. But such far-sighted minds as Democritus's were comparatively rare.

Following the thinking Greeks, society chose to hibernate mentally during the Dark Ages. In the course of these wasted centuries everything was done to exaggerate the importance of man, and to minimise the age and size of the universe. And then, with the advent of the Renaissance, scientific discoveries destroyed man's geocentric throne. The universe became enormously greater, and man was cut down to his proper size.

The Theory of Evolution, by which Darwin demonstrated that life-forms had been changing and improving in the course of millions of years, dealt a blow to the greatest fallacy of man—anthropocentrism. It showed that man and the world were much older than had been believed, and that biologically man was a continuation of the animal. The dialectical materialism of Heraclitus and Democritus which claimed that Nature is all interconnected, fell into place on the background of these scientific revelations.

'When he hears of Greece, the cultivated German finds himself at home,' wrote Hegel. His dialectics of the contradiction of polarities laid the foundation for the work of Engels. Thus the philosophy of Heraclitus was carried through the nineteenth and into the twentieth century. But the germs of dialectical philosophy had existed from the very dawn of civilisation. The ancient Greek, Buddhist, Brahmin and Chinese thinkers professed belief in a world where nothing is permanent except the law of change.

Krishna expresses this doctrine in the Bhagavat Gita:

This vast company of living things,
Again and yet again produced, expires
At Brahma's Nightfall, and at Brahma's Dawn,
Riseth without its will to life new-born.

This is an excellent picture of the appearance and disappearance of worlds in cosmic space, which continues without end. After his enlightenment Buddha said: 'I believe that the world is going to exist forever and forever. It will never come to an end. And anything that has no end, has no beginning. The world was not created by anyone. The world always was.' The Buddhists do not believe in a creator. Neither did the ancient Greeks who held that it was the universe that created the gods and not the other way around. The *Rig Veda* confirms that a similar belief existed in ancient India:

> The gods are later than this world's formation;
> Who then can know the origin of the world?

From the earliest times the Taoists of China postulated the existence of two principles in Nature—the positive Yan and the negative Yin, the interplay of which produces life and motion. The dialectical philosophy of Heraclitus recognised the opposing faces of Nature and the law of constant change. It did not regard Nature as an accidental collection of things and phenomena but as an integral whole in which everything was interlinked.

This school of thought taught that all progress and evolution is an onward and upward movement, a transition from an old qualitative state to a new one, from the simple to the complex, from the lower to the higher. According to Diamat philosophy, mind is the by-product of matter. This does not really downgrade man for, after all, has not man been able to manufacture material particles in atomic accelerators? Perhaps the argument which came first—mind or matter? should not be posed as sharply as—which came first, the chicken or the egg?

In this century science discovered the fire of the atom. Einstein made his revolutionary discoveries about motion, mass, energy and time, thus exemplifying a new model of the old universe. In the last 15 years, for the first time in history, man has learned to fly in space, bridging the gaps to the moon, Mars and Venus. Never before has man acquired so much knowledge in so little time.

Yet regrettably the mass psychology of twentieth-century man is generally similar to the mentality of a shopkeeper of the sixteenth century or a peasant of the fourteenth century, despite his having cars and television sets. The historian of tomorrow will be appalled at the abyss separating the new knowledge from the outdated beliefs inherited from the Dark Ages. This disparity is one of the principal reasons for our present material and psychological crises. One can not use eighteenth-century politics in the Atomic Age. One can not adhere to old economics when all nations are already linked by trade and finance. One can not continue to cherish earthbound superstitions in this era of space exploration.

Scientific philosophy is a dialogue from which can emerge the truth. Dogmatic religions or restricted ideologies are monologues in which only one party does the talking—the one who thinks he has a monopoly on truth, even if he is wrong.

Oh let us never, never doubt
What nobody is sure about,

wrote Hilaire Belloc.

When religion is a personal experience of infinity and eternity or when it constitutes an attempt towards philosophical understanding, it is a perfectly natural, beautiful and healthy phenomenon. But when it dominates politics and is incorporated in a state's structure, then can rise the sword of Constantine and the firebrands of the Inquisition, destroying that which it was supposed to build—brotherhood.

However, religion can and should make men brothers. In Lebanon I have seen a church near Beirut divided into two sections—Roman Catholic and Greek Orthodox congregations co-existing peacefully side by side. There are other temples in that land, similarly partitioned, which accommodate Christians in one part of an edifice and Mohammedans in the other. If such tolerance is possible in Lebanon why can it not be imitated in other, less cultured, parts of the world?

Oppression and exploitation breed revolutions. But when, after victory, the revolutionaries monopolise the ideal of social justice, when power is held by a very few, when judgment, good or bad, is exercised by a very few, what happens? Human beings are turned into zombies, the walking dead.

Again the same outrage—the song is good but the cursed singers are chanting falsetto! Intellectual splendour can appear only in a free society in which ideas are freely presented, accepted or rejected by people.

The erring establishments, political, religious and scientific, which have sold mankind defective merchandise at costly prices, should only be laughed at. This was the bidding of Democritus—to laugh at the folly of mankind. The mental blocks of masses reveal the magnitude of a genius. Sceptics and diehards pave the way for a pioneer of thought, just as a flashlight shines in the dark.

The gloomy night of philosophy and science which covered the Western World in the Middle Ages is not completely over yet. At his trial Socrates was accused of being an evil-doer for 'not worshipping the gods the state worshipped'. This practice by the establishment of crushing free independent thought, for fear it would disclose the falsity of the worshipped idols, has persisted for centuries. The establishment, incapable of proving the deficiency of the logic of its critics, finds a simpler solution—that of ostracising them.

The world is full of stereotyped people who repeat only the words of ideological programming centres, some too lazy, some afraid and some incapable of thinking for themselves, Detestable is the sight of such human-like robots. It is awful to see how whole portions of the human ocean have been programmed by invisible stations broadcasting on a mental frequency that reduces men to the parrot status.

One does not prove the truth of dogma by burning a man at the stake or sending him to a concentration camp. Persecution discloses the inability of the establishment to defend its tenets by logic and philosophy.

Human nature is so rich; why should there be mass-produced citizens? Has this not brought enough bloodshed for one planet? Let a thousand schools of thought meet and discuss as they did in ancient Greece.

Mankind's intellectual maturity grows as the barriers of space and time are progressively broken. When new continents were discovered in the Era of Discovery, the consciousness and way of life of the Europeans changed radically. When archaeology uncovered the ruins of Egypt and dated that civilisation's source to a time when, according to prevalent ecclesiastical opinion, the world had not yet been created, the origin of civilisation had to be moved back for thousands of years. These crumbling frontiers of space and time opened new horizons.

The aim of this introductory section, THE TREE OF LIFE, has been to show the growth of intelligence and consciousness from the infinitely small units of life to greater cosmic systems. This was a vertical view of evolution from its foundation to pinnacles lost in the stars.

The next part, ON THE SHORES OF ENDLESS WORLDS, is devoted to the examination of the possibility of life around us in cosmic space on a level more or less approaching our own. If life is potentially inherent within the core of the atom, it must sooner or later manifest itself when the environment is favourable. It must blossom somewhere in the galactic systems in space with their billions of suns and planets, as the chemical elements in the furthest reaches of our galaxy are identical to those on Earth.

PART II ON THE SHORES OF ENDLESS WORLDS

11: *Islanders in Space*

On the seashore of endless worlds children meet,
They build their houses with sand and
They play with empty shells.

Rabindranath Tagore presented his vision of our astronomical location in the universe with this poem. The 'seashore of endless worlds' is, of course, the outer fringe of the Milky Way galaxy where our solar system hovers. There are more than 100,000 million suns in this vast sea of stars. Our sun, a very ordinary star, with its family of nine planets, thirty-one satellites and countless comets, form a particle of this enormous swarm of stars, which revolves around its axis once in about 200 million years, travelling through space at the same time.

The planet Earth is an island orbiting the sun. It is here that we 'build our houses with sand and play with empty shells'. Our astronomical address is as follows: we are natives of a space island, possessed of a biosphere, with an approximate diameter of 12,700 kilometres, of an estimated age of 4,500 million years, located in the peripheral zone of the galaxy.

How would our solar system appear to astronauts in a spaceship at a distance of 300 light years? The sun would look like an orange star of apparent magnitude 7 with a surface temperature of about 11,000 degrees centigrade. They would be unaware of the existence of giant Jupiter, to say nothing of the small Earth.

For many centuries we did not know that we were islanders in space. When this realisation came, we began to wonder if there were others like us. Without comparisons it would be impossible to understand the immensity of the Milky Way galaxy of which we are part. Our Earth is no bigger than a seed floating on the waves of the Pacific, compared with the size of our galaxy. Our solar system is separated from others by incredible depths of space. It takes light rays from the nearest star Proxima Centauri over four years to reach Earth yet light flies around the Earth seven times per second.

Planets, stars and galaxies have life-spans of milliards of our years. In the grandiose theatre of the telescope astronomers observe the births and deaths of worlds. Like a forest composed of young, mature and old trees, are the stars in our galaxy. The chemical elements of our solar system can be found at the furthest reaches of the Milky Way since matter is the same in substance throughout the universe. By the same token, natural laws, forces such as gravity, electricity and nuclear energy, are also identical everywhere.

Over a century ago Victor Hugo wrote: 'Is it not imaginable that during the probable existence of billions and billions of centuries, the myriads of stars and suns, though obeying the universal laws of birth and death, undoubtedly have beginnings and ends, transforming themselves, replacing themselves and renewing themselves unceasingly, without respite, without end, forever and ever and ever.'[1]

Young stars are generally of a brilliant blue colour, mature ones are white, and when they reach a yellow hue they have become middle-aged, as is our own sun. When an astronomer sees one that is blood-red, he knows that before his vision is a dying star. Taking into account the immense number of suns in our galaxy, it would seem that there are millions of stars similar to our sun. Some astronomers calculate that one star out of four possesses a planetary system.

Owing to astronomical distances it is impossible to see these

[1] *Choses de l'infini dans Post Scriptum de ma vie*, 1864 (author's translation).

planets from the Earth; however, the existence of some of them has already been proven by the irregular motions of stars. On the basis of these perturbations a number of dark companions, invisible planets, have been found within a radius of 10 light years from our solar system. Proxima Centauri has one large planet, Barnard's star three planets about the size of Jupiter or Saturn, Lalande 21185 one giant companion, Ross 614 a superplanet. An examination of 240 of the nearer stars has disclosed that 60 of them show periodic perturbations which could be explained by the presence of planets.

In observing the ring of Planetary Nebula NGC 1501, astronomers are now actually watching the formation of a planet or planets near a crystallising star. It has also been noticed that stars of the yellow–white type, like our sun, rotate slowly on their axes. This low angular momentum might be attributed to a drag caused by the weight of planets. This is why the stars Tau Ceti, Epsilon Eridani and Epsilon Indi are suspected of having planetary systems.

Planets in the habitable zone where the stellar radiation is neither too intense, nor too weak, can produce life. They must, of course, have chemical components of proteins. As Isaac Asimov writes in *20th Century Discovery*, 'any planet that is something like the Earth, with a nearby sun and a supply of water and an atmosphere full of hydrogen compounds, would then have to form life'. The infinite variety of shapes and compositions of living organisms on Earth compels us to believe that life-forms in other worlds would be just as diversified in their structure in order to adapt themselves to their environment.

In his classic work *Dialectics of Nature*, Friedrich Engels writes: 'When organic life becomes a fact, then it must evolve to the stage of rational beings in the course of generations.' In another passage of the same book he says: 'And although iron necessity will force it (matter, A.T.) at some time to destroy its sublime blossom-sentient spirit, it will have to revive it again in some other place at another season.'

Today many of the world's foremost scientists are wholly in agreement with this conclusion. The noted British biologist J. D. Bernal has said: 'We have no convincing reasons for considering that we are unique—life may and indeed must have occurred in other parts of the universe.'[2] Another distinguished British scientist, Richard Calder, thus estimates the number of planets in our galaxy: 'That would give us billions of possible earths and, with what some call "mathematical certainty", millions of worlds at precisely the same stage of development as our own.'[3]

The American astronomer, the late Harlow Shapley, stated: 'Life is widespread. It evolves out of the lifeless as a natural product of cosmic evolution.'[4] Dr Thomas O. Paine, former director of the National Aeronautics and Space Adminstration, stated in London on a BBC 1 television programme on 31 January, 1971: 'It is inconceivable that in a galaxy of millions of stars we should be alone.'

Yet the French biologist, Jacques Monod, writes in *Chance and Necessity:* 'Man at last knows that he is alone in the unfeeling immensity of the universe out of which he emerged only by chance.' This dogmatic statement can be questioned on mathematical grounds alone, considering the millions of stars and planets in the galaxy. Nevertheless, Professor Monod's conclusions concerning the living world as 'the product of a gigantic lottery that draws numbers out at random' are accurate. Nature's Monte Carlo game to which Monod refers is the very agent which can bestow life upon a given world. Since the roulette of life is played in millions of galaxies during billions of years, the odds are in favour of there being a very large

[2] J. D. Bernal, *The Origin of Life*, London, 1967.
[3] R. Calder, *Man and the Cosmos*, London, 1968.
[4] H. Shapley, *Beyond the Observatory*, New York, 1967.

number of winners, to say the least. In our galaxy alone, solar systems of the sun type must exist by the millions. These galactic casinos were open long before the birth of our sun and the gambling must be continuing non-stop. The mathematical probability of winning tickets, with life as the prize, must be enormous, beyond all expectations.

Professor Jacques Monod completely ignores the thesis of dialectical materialism that all energy, life and even mind might be contained in the core of the atom, ready to sprout and give fruit whenever conditions are suitable. His view of the aimlessness of all life is shared neither by dialectical materialists nor by religionists. Speaking of chance, Engels wrote: 'The accidental is necessary, and the necessary is also accidental.'

Monod's assertion that 'man is alone in the universe' revives the medieval notion of the exclusiveness of man, so out of place in a man of science. However, the debate concerning whether life is the exception or the rule will be quickly closed when only a tiny piece of moss or a fossil of some life-form is found on Mars or elsewhere.

Meanwhile, scientists armed with telescopes rather than microscopes, like Jacques Monod, are reaching totally different conclusions to his. Australian radio astronomers, using the giant radio-telescope at Parkes, have detected the organic chemical formaldemine in deep space. Commenting on this discovery, Professor R. D. Brown of Monash University remarked: 'This suggests that carbon-based life could easily evolve in other parts of the galaxy.'[5]

A prominent American spectroscopist, Dr Fred H. Johnson, when analysing spectral lines taken of the Milky Way by the large optical telescope at California's Lick Observatory, discovered TBP (magnesium tetrabenzporphine)—a forerunner of chlorophyll or the green plant pigment that triggers the process of photosynthesis on Earth. 'I believe this is one of the

[5] *Daily Telegraph* (London), 31 May, 1972.

first clues of the building blocks of life in the universe,' Dr Johnson stated.[6]

Radio astronomers have already detected in interstellar space such compounds as water, ammonia, formaldehyde, acetylene, hydrogen cyanide and carbon monoxide—an excellent assortment of chemicals from which to produce life. Space probes have disclosed that the moon may have water below its surface. Mars is covered with dried river-beds and traces of water vapour are present in the Martian atmosphere. So far, Mars offers the most likely possibility of discovering some sort of rudimentary plant or bacterial life.

A person with average sight can see only a few thousand stars in the sky. With a telescope he is able to observe millions. A portion of these suns with planets could have generated plant, animal and human life at different stages of development.

After having studied ancient Greek writings Giordano Bruno concluded that: 'There are innumerable suns and innumerable earths which revolve around their suns as our seven planets revolve around our sun. These worlds are inhabited by living creatures.' Yet another example of one who said the right thing at the wrong time, for as a result he was burned alive at Campo dei Fiori in Rome in 1600.

This conception was further developed a century later by the Dutch astronomer Christian Huygens whose book *The Celestial Worlds Discovered* was published in London in 1698. In this work he expressed his conviction that there were 'an innumerable company of worlds in the stars' and that their inhabitants 'view the same stars as we'—magnificent thoughts, indeed. Huygens even drew their portraits, employing mathematical logic and natural history. The dweller of another planet is probably not much larger than we are. The limiting factor here is something called the square-cube law and gravity. If you double the height of a person without changing his or

[6] *Herald-Tribune* (Paris), 28 October, 1971.

her proportions, you then have a being eight times the original weight. For example, if a man were 1·75 metres tall and weighed about 65 kilogrammes and we doubled his height without changing his proportions, he would weigh 520 kilogrammes on a planet about the size of the Earth. He would, therefore, be much too clumsy to work and would be unable to survive.

The hypothetical intelligent being of another world would be better off possessing two eyes as does man and all creatures on Earth because focusing permits the judging of distance and shape. He must have a soft brain with a hard skull to protect it. These are the elementary rules of *biological construction* which were calculated by Huygens. With a few amendments they are still valid today.

The shape and physiology of the human body are dependent upon the geographical and astronomical determinants. Thus swarthy Aryans coming to the north of Europe developed pale skins and blond hair because of the lack of malanin due to the weaker and scarcer sunshine. The Quiche Indians of Bolivia and Peru are able to live at an altitude of 4,500 metres after having developed voluminous lungs for oxygen intake. The heartbeat of the Maya in Yucatan's oppressive jungle heat is about 52 pulsations a minute, compared with the 72 beats per minute of the European.

On another world the planetary determinants would mould plants, animals and rational beings in such ways as to make their appearance totally different from that of terrestrial forms.

Since H. G. Wells's *War of the Worlds* a large portion of Anglo-American science fiction is geared to the idea of monsters from space. However, these ideas do not appear in the science fiction of Eastern Europe or of the Orient, though there is a small amount in Japan and the Latin nations. The belief that we are angels on Earth attacked by devils from the heavens would seem to indicate a schizophrenic complex on a mass scale. It is devoid of logic and is a manifestation of the egocentric, geocentric and anthropocentric attitude of Western man. Moreover, it is now quite obvious that our civilisation is perfectly capable of completely destroying itself without any outside help from space monsters.

A Russian science fiction reviewer, N. Petrov, writes: 'Soviet science fiction writers are full of the optimistic belief that intelligence, as opposed to insanity, is a blessing. But if reason turns into madness, which is not impossible, it then destroys itself. The higher the stage of intelligence the wider the cosmic sense of humanism emerging from it.'

Wildly imaginative tales of lands outside the known world, beyond the reach of salvation and Christian grace, are found in the book of Sir John Mandeville, a thirteenth-century geographer. They are reminiscent of modern notions of space monsters. Men without heads with their faces on their chests or men with dog heads, one-legged men with a gigantic foot—all these were drawn by Mandeville. These monsters were naturally regarded as the offspring of the devil, not having been created in the image of God like Sir John.

A thousand years before Mandeville's outlandish devils, Plotinus was more scientific in his speculations: 'Nor is Earth alone adorned with an endless variety of plants and animals. The celestial spaces are filled with illustrious souls.'

It is possible that owing to such factors as atmospheric conditions, gravity and unimaginable foods as well as evolutionary ancestry, planetarians elsewhere might appear very odd to us. In fact, by our standards built on the prototype of a nicely dressed super-ape, dwellers of other worlds may even appear quite weird to us. This feeling would probably be shared by our cosmic kin should we meet.

Hopefully, the form of these hypothetical citizens of stellar empires would not be too disturbing to us. Rather we should consider the level of consciousness, intelligence and ethical behaviour that these beings have attained. Some may be our equals, whilst others our juniors, and there may be planetary dwellers in certain systems whose appearance and mentality will be strikingly superior to ours. A greatly older galactic civilisation may have attained the highest degree of both knowledge and awareness. A number of planetary societies may have already passed through the painful stage that we are now traversing. Others could have already established one-planet economies and amalgamated themselves into stellar federations.

If such theorising is logical and scientific, then somewhere along the line we may find traces of the presence of our cosmic relatives. So let us now take a close look at the starry sky, as if we have never seen it before, and repeat the words of Enrico Fermi: 'Where is everybody?'

12: *We Shall See and Hear Other Worlds*

Earthman has already trodden lunar soil. He has constructed equipment by means of which he can see our nearest neighbouring planets—Mars and Venus. For information further afield Pioneer 10's mission was to bypass Jupiter and then leave the solar system entirely.

In the seventies NASA, with the co-operation of six nations, is planning to launch an unmanned spaceship to tour the outer planets. Eventually rockets powered by a combination of chemical fuel, nuclear and electrical engines will explore the solar system.

However, in order to reach distant stars our astronautics will have to develop a far more rapid and very different form of propulsion in view of the thousands of years required by today's methods to cover distances to other solar systems.

Professor John A. Wheeler of Princeton University has proposed a most intriguing theory—that of instant space travel through the region of what he defines as *Superspace*. He compares the shape of cosmic galactic systems with a doughnut where there is no time or space. According to this theory, a spaceship leaving the body of this astronomical doughnut and flying through the hole, would take no time at all to reach a star world across that Superspace. As in this Superspace *before*, *after* and *next* would have no meaning, the astronauts would be transported instantaneously. But it will probably be many centuries before science is capable of putting this theory to practice.

In the context of Einsteinian physics it is impossible for any material object to travel faster than light as the mass of the object would disintegrate. However, a rocket flying on a photon ray at a speed a fraction slower than that of light may in the future be the fastest method of crossing interstellar space.

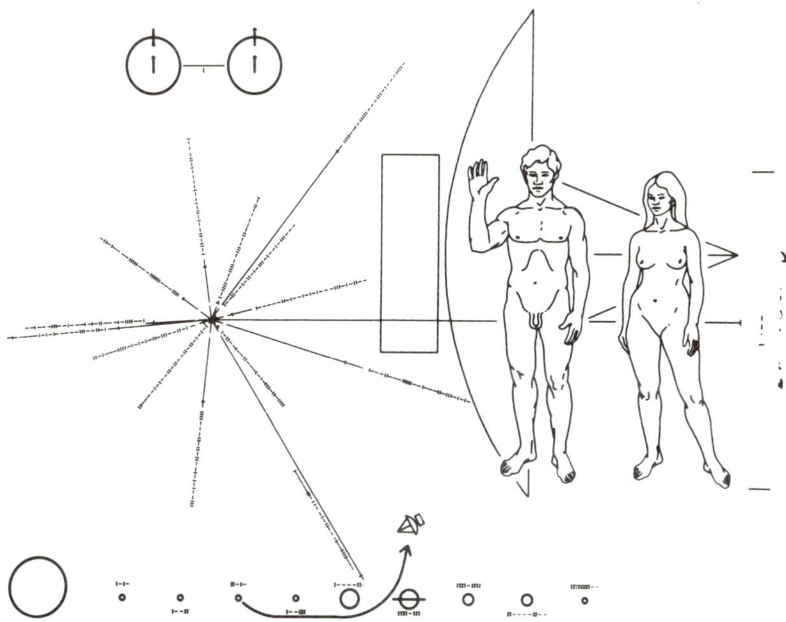

Pioneer 10 space probe carries a symbolic message from the people of Earth to extraterrestrial beings.

Blueprints of photon ray, antimatter and electrothermal (ionic and plasma) engines already exist but their construction is another matter involving many technical problems which might require decades to solve. Nevertheless, in principle, an antimatter starship is certainly a future possibility. It would have a very strange appearance—a combination of the Jodrell Bank radio telescope laid flat on the ground with the Eiffel Tower placed on top of it. The concave surface of the portion resembling the radio telescope would have to be covered with a mirror of maximum reflecting power. From the centre of this enormous bowl would flow a stream of photons, creating a gigantic jet, a pillar of light comparable only to the tail of a comet, which would push the rocket towards the stars. The photons would be the result of the annihilation of matter and antimatter, so controlled as to provide a steady stream. When an electron collides with an anti-electron (positron) they annihilate each other, producing two photons, thus converting matter into light, photons from both matter and antimatter being identical.

The main technical problems would be, first that of manufacturing sufficient antimatter fuel in the atomic accelerators, and second that of isolating antimatter from matter in the propulsion system to prevent the spaceship from exploding. Electromagnetism and plasma are expected to furnish such an isolation layer. The interstellar rocket would have to be assembled in space, and then launched at a tremendous distance from Earth as the gigantic jet of light from its concave mirror would be capable of reducing our oceans to steam.

An ordinary unmanned rocket of current usage would take around 50,000 years to reach Proxima Centauri, the nearest star, but a photon rocket would slash the travel time to a minimum.

According to Einstein's law of time shrinkage, when a spaceship flies at a speed approaching that of light, the flow of time is reduced. Let us suppose that its speed is 299,000 kilometres per second or just one kilometre below the speed of light. The chronometer and log book of such a spaceship would record, upon its return, that the trip lasted one week. Yet during this week on shipboard, the Earth would have turned around the sun $8\frac{1}{2}$ times, that is $8\frac{1}{2}$ years would have elapsed for the people of our planet.

What really would occur is this—when the mass of a starship approaches the speed of light in its flight, all atomic oscillation would slow down. Everything would decelerate uniformly—the ticking of the clock, heartbeats, breathing, digestion and the very act of living, so for those aboard the ship, though unaware of the phenomenon, time would crawl as compared with earth time. This time contraction can be more successfully understood by observing the movement of a hand on the face of a clock. Near the centre the hand travels along short arcs but on the edge the arcs are many times longer. Our imaginary spaceship would be somewhere near the centre of the cosmic clock, and what would seem to be just a few days to those aboard, would be $8\frac{1}{2}$ years to the people on Earth.

The Special Theory of Relativity has now been proven beyond doubt by two American physicists, Joseph Hafele and Richard Keating. To test the theory they twice flew four atomic clocks around the world in jet aircraft—the first trip eastward and the second westward. On their return to Washington, D.C., they found that their clocks were a hundred billionth* of a second behind the Washington set. Had they stayed at home, they would have aged one billionth of a second more.

Interstellar photon ships are not likely to be hurled into space in the near future; however, it is possible that owing to the explosive growth of our technology and science, the project might become feasible in the next century.

Meanwhile, a cone-like space projectile has been sent to infinity, carrying a message from the people of Earth to any extraterrestrials it may encounter. Pioneer 10 was launched in March, 1972 towards the planet Jupiter which it was to bypass, then proceed beyond the solar system. It will travel for ever unless intercepted by a spaceship of another world. For such a contingency the space-cone contains an alluminium plaque anodised with gold, on which the American astronomers Carl Sagan and Frank Drake draw a chart with the help of the artist Linda Sagan.

This plaque depicts an atom of hydrogen undergoing a

* American billion—milliard in Europe.

change of energy state, a starburst pattern symbolising 14 pulsars, the distance of the sun from the centre of our galaxy, and a diagram of the solar system showing that the spaceship left the third planet and was headed towards Jupiter. A sketch of man and woman—the planetary dwellers responsible for this sophisticated cosmic message, is also included.

It is presumed that any civilisation possessing the intelligence and the technical means of recovering this space-cone with the plaque would surely have the prerequisities for decoding its message.

The theory of the existence of superior galactic civilisations, the rate of development of these civilisations and the methods of contacting them was first proposed by scientists Morrison, Cocconi and Bracewell in 1959–60.

Our sun is estimated to be 5,000 million years old. Since the age of the Milky Way galaxy is placed somewhere between 10,000 and 25,000 million years, it would seem reasonable to assume that there must exist millions of solar systems immensely older than our own. A galactic civilisation 100 million years older than that of Earth could have reached undreamed-of heights in culture, science and technology. If such a still existent civilisation had ever been warlike at any time in its history, its conflicts would undoubtedly have been finally resolved and a planetary state created, otherwise it would have, no doubt, destroyed itself by means of a technology far in advance of its capacity for moral judgment.

The physical laws we know are not confined to Earth but are extant throughout in the universe. A super-civilisation may possess greater knowledge concerning physics and astronomy, yet whatever we know ourselves, they also must know. Accordingly, there do exist some mutual areas of comprehension upon which understanding could be achieved.

It has been suggested in the science fiction novel *Andromeda* by the Soviet scientist and author Ivan Efremov that extraterrestrial intelligences form federations at an advanced cultural and technological stage. He calls it the *Great Ring* of civilisations. Apparently, we are still outside this privileged circle.

On our own planet humanism has always been a mark of culture and civilisation. It is only logical to assume that a planetary society with an advanced technology would either create a world built on the ideas of peace and progress, or else destroy itself.

In his article in *Nature* (28 May, 1960) Professor R. N. Bracewell wrote about such a galactic federation: 'There may be some communities that have achieved durability, even quasi-permanence, perhaps gaining control of the circumstances that lead to short average life-times. Aided by accidental proximity due to random spacing, some of these could be in contact. Presumably such an ancient association would be very able indeed technically, and might seek us out by special means that we can not guess. Whether they would be interested in rudimentary societies which in their experience would usually have burnt themselves out before they could be located and reached, is hard to say.'

The British biologist J. D. Bernal has some far-sighted views on the subject: 'There is a possibility that the oldest and most advanced civilisations on distant stars have in fact reached the level of permanent intercommunication and have formed as it were, a club of communicating intellects of which we have only just qualified for membership and are probably now having our credentials examined.'[1]

The celebrated British physicist Eddington said a few decades ago: 'Somewhere in the universe there may be other beings a little lower than the angels, whom man may regard his equals, or perhaps his superiors.'[2]

If we wanted to advise our galactic neighbours that we think

[1] J. D. Bernal, *The Origin of Life*, London, 1967.

[2] A. S. Eddington, *Space, Time and Gravitation*, Cambridge, 1920.

we are now mature enough to be included in the Universal Telephone Directory, how could we do it? The experts say—by means of radio in the 1,000 megacycle region, or by optical light using coherent laser light in a blacked out line of a common element, or else by energy gamma rays.

There may be two other possible methods—the use of neutrino particles or of telepathy, neither of which are affected by electromagnetic forces. Our knowledge concerning both of these fields is still very limited, and yet these may be the methods by which galactic civilisations communicate with each other.

However, at the present time the most promising procedure would appear to be that of a radio watch on the 21-centimetre wave or 1,420 megaherz which corresponds to the line of hydrogen. This was already utilised in 1960 when the astronomer Frank Drake tuned in on the stars Epsilon Eridani and Tau Ceti which are situated at the reasonably close distance of 11 light years from us. This was the experiment known as the *Ozma Project*, the results of which, unfortunately, were disappointing. Nevertheless, astronomers refuse to be discouraged and, at present, the Radio-Physical Institute at Gorky, U.S.S.R., is conducting a search on 50 selected stars in our part of the galaxy. When America's 300-metre radio telescope, built in the crater of an extinct volcano at Arecibo, Puerto Rico, is modified, it should be able to capture radio signals emanating from a distance of 10,000 light years. In the northern Caucasus on Mount Semirodniki the Russians have constructed the world's biggest optical telescope with a diameter of 10 metres. It operates in conjunction with the 600-metre radio telescope near by, known as RATAN-600 to capture faint signals from distant stars.

Earthman unknowingly joined the network of galactic radio stations when, earlier this century, he began his first wireless transmissions. Our radio waves continue travelling through space until they are picked up by some technological civilisation of a distant planet, thereby arousing interest as to the source of these emissions. This could be compared with the effect of throwing stones into a pond and creating concentric ripples.

Dr Charles Townes of the University of California, a Nobel laureate, has suggested the use of lasers (light amplification by stimulated emission of radiation)—a super-powerful light beam projectors which could actually send from Earth a piercing light ray that would be detectable on a planet 10 light years away. Then he plans to examine the spectographs of the stars aimed at. If any of the Fraunhofer lines in the known spectroscopic patterns of a particular star were to show a sharpening or brightening effect, it might be a super-laser response signal beamed to us from another planetary system.

Dr Townes has also said that automated space probes should be sent to the stars. They could televise pictures upon their arrival giving information regarding the world from which they came. However, this could only be carried out if the civilisation sending these probes is a permanent one as with our present technology it would require thousands of years for them to reach their destinations. Astronomically, mankind on this planet is still comparatively young and the permanence of its civilisation most problematical.

In September, 1971 an international conference was held in Soviet Armenia at the Astrophysical Observatory of Byurakan to consider the problem of communication with Extra-Terrestrial Intelligence, abbreviated as CETI. Among the delegates were such celebrated scientists as Dr Carl Sagan (U.S.A.), Dr Joseph Shklovsky (U.S.S.R.), Dr Frank Drake (U.S.A.), Dr Victor Ambartsumian (U.S.S.R.), Dr Francis Crick (U.K.), Dr Freeman Dyson (U.S.A. and U.K.), Prof. V. Ginzburg (U.S.S.R.), Dr Charles Townes (U.S.A.) and many others.

At this convention it was agreed that there must exist galactic civilisations which have probably developed along lines similar to those of earthman. There was wide agreement on the possibility that a number of extraterrestrial cultures could have attained extraordinary capabilities.

A resolution was passed recommending the setting up of a project to intercept signals between these stellar communities. A permanent international group was appointed to co-ordinate research in this field. Two areas of exploration were proposed: one section to concentrate on discerning whether or not any of the several hundred stars under observation were being influenced by super-civilisations which might be modifying their solar systems by means of *astro-engineering*, the other assignment to seek the beacons of remote civilisations including those in nearby galaxies, millions of light years away.

The host of the conference, Professor Ambartsumian, recapitulated the CETI convention in these words: 'All that we know can be summed up in one sentence—life and mind have appeared on other planets provided the environment was suitable. Therefore, in principle, contacts with rational inhabitants of other worlds are possible.'

Radio telescopes are scanning the skies for strange radio signals yet doubt clouds our minds because of the time and space scales involved. But there have been men who believed in the reality of our cosmic antipodes without seeing or hearing them.

I hear beyond the range of sound,
I see beyond the range of sight,
New earths and skies and seas around.

This was written by the American essayist Henry Thoreau over 100 years ago.

Nikola Tesla, the first inventor to propose an effective way to use an alternating current, believed he had received a message from space. In his laboratory in Colorado Springs he wrote in his diary in 1899: 'The changes I noted were taking place periodically, and with such a clear suggestion of number and order that they were not traceable to any cause then known to me. It was some time afterwards when the thought flashed upon my mind that the disturbances I had observed might be due to intelligent control. The feeling is constantly growing on me that I had been the first to hear the greeting of one planet to another.'

Guglielmo Marconi, the celebrated pioneer in wireless, had a yacht in the Mediterranean with a large transmitter-receiver. In September, 1921 Marconi intercepted signals on the 150,000-metre wavelength. The maximum wavelength utilised at the time was only 14,000 metres. The regularity of the signals disposed of any assumption that the waves might have been caused by electrical disturbance. The signals were unintelligible, consisting apparently of a code. Marconi was convinced that the signals came 'from some place elsewhere in the solar system'.

When Mars made its close approach to Earth in 1924, there was considerable speculation at the time as to whether the planet was the abode of some type of intelligent life that could have sent radio messages to our world. The *New York Times* duly reported that: 'C. Francis Jenkins who has been a prominent experimenter in the field of transmitting photographs by radio, set up a machine designed to receive on a moving film any message that might come splashing through the ether.'[3]

Plans for the experiment had been carefully made by Dr David Todd, Professor of Astronomy at Amherst College, Massachusetts. On the astronomer's initiative all countries with high power transmitters were requested through diplomatic channels to silence their stations for five minutes every hour from 11.50 p.m. 21 August to 11.50 p.m. 23 August, 1924.[4]

[3] *New York Times*, 23 August, 1924.

[4] W. Sullivan, *We Are Not Alone*, London, 1970.

The apparatus, constructed by C. Francis Jenkins of Washington, D.C., was attached to a receiving set adjusted to a wavelength of 6,000 metres. In this instrument the film tape was 9 metres long and 15 centimetres wide and the incoming radio signals were converted into flashes of light, immediately printed on the film. The device was switched on at the specified times and when the film was developed four days later, Professor Todd and C. Francis Jenkins found a fairly regular succession of dots and dashes along one edge. There have also been reports that the film showed repeated images of something that appeared to be a human face. Concerning this extraordinary phenomenon Jenkins was quoted as saying: 'It's a freak which we can't explain.' Unidentified dots and dashes were also received by R. I. Potelle, Chief Engineer of Station WOR, Newark, N.J., on the night of 22 August, 1924. The signals were steadily repeated but could not be decoded.

Equally curious reports appeared later in the 1920s. One came in 1928 from Professor Carl Störmer, the noted Scandinavian researcher on the Aurora Borealis. He discovered that radio signals sent out on the 31·4-metre band by the powerful short-wave station of the Philips Laboratory at Eindhoven, Netherlands, returned as radio echoes with a delay of a few seconds. Radio waves travel 300,000 kilometres a second, which means that if radio or radar is beamed at the moon, the waves will hit the moon in about 1¼ seconds and the echo will return in approximately 2½ seconds from the time of transmission. A radio echo from Venus would come back in about five minutes and one from Mars in nine minutes.

From the Eindhoven transmissions which were received by Störmer in Norway, it can now be surmised that the reflecting objects in space were situated just beyond the moon but not as far as Venus or Mars.

On 11 October, 1928 Professor Störmer recorded long-delayed echoes (LDEs) with the following time delays between broadcasts and receptions: 15, 9, 4, 8, 13, 8, 12, 10, 9, 5, 8, 7, 6 seconds, and 12, 14, 14, 12, 8, 12, 5, 8, 12, 8, 14, 14, 15, 12, 7, 5, 5, 13, 8, 8, 8, 13, 9, 10, 7, 14, 6, 9, 5, 9 seconds. The distances of the astronomical bodies which could have reflected these signals can easily be calculated. The longest interval was 15 seconds of 7½ seconds each way. That body should, therefore, have been at a distance of about 2,250,000 kilometres from the Earth. However, there is no known asteroid or planet in that orbit. The shortest interval was 3 seconds, or 1½ seconds each way, which places the reflecting object just beyond the moon.

On that same evening—11 October, 1928—Carl Störmer in Oslo sent a telegram to Dr B. van der Pol in Eindhoven with a request to transmit test signals. During the night his receiver picked up echoes with the following lag—8, 11, 15, 8, 13, 3, 8, 8, 8, 12, 15, 13, 8, 8 seconds. Apparently the phenomenon was most irregular as Dr van der Pol wrote later to *Nature* magazine[5] stating that his radio station 'did not hear any of these long period echoes for several months after'.

Although attempts have been made to explain the Störmer–van der Pol returned signals as reflections from ionised gas scattered around the solar system as well as magnetic storms or atmospheric quirks—these interpretations have not solved the mystery of the echoes.

Then in 1960, Professor Ronald N. Bracewell of Stanford University attributed the long-delayed echoes to an interstellar computorised probe from another world in space, which had been listening to our signals and repeating them at intervals. 'Should we be surprised if the beginning of its message were a TV image of a constellation?' asked Bracewell.

Taking this hint early in 1973 the young Scottish astronomer and scholar Duncan A. Lunan proposed a sensational theory. He plotted a graph of one sequence of the long-delayed

[5] *Nature*, 8 December, 1928, p. 878.

echoes, recorded in 1928, by showing delays in seconds horizontally and pulse sequence vertically. A pattern of dots immediately emerged which displayed a remarkable resemblance to the constellation of Boötes, the Herdsman.

It would be appropriate to state here that the idea of the conversion of radio signals into a graphic message is being planned for our future cosmograms to other worlds in space. Evidently, the interstellar probe orbiting in the vicinity of our planet had already employed this technique in 1928 which indicates the technological superiority of the civilisation that had launched it.

Duncan Lunan has traced the source of this galactic satellite to Epsilon Boötes, a double-star system 103 light years away. In the graph Arcturus, a bright star within this constellation, appears displaced from its position today. Owing to this peculiarity Lunan was able to calculate the date of the probe's arrival to approximately 11,000 B.C.

In one particular series of delays the 8-second echoes seemed to draw a barrier across the chart, to the right of which was a series of dots clearly reminiscent of the configuration of the constellation of Boötes. Similar graphs by Lunan, based on other LDEs, produce recognisable maps of the constellations of the northern hemisphere. Long-delayed echoes were also mentioned by a French astronomical expedition to Indo-China in May, 1929. According to Galle and Talon who wrote papers about this phenomenon, the delay times ranged from 1 second to 30 seconds. On the basis of their reports Duncan Lunan constructed other charts with dots and then, by connecting them in a certain way, a geometrical pattern emerged, which he decoded and interpreted as follows:

AB—Start here
BC—Our home is Epsilon Boötes
CDE—which is a double star

FG, GH—We live on the 6th planet of 7
CH, GK, JKL—check that, the 6th of 7
EM—counting outwards from the sun
FEG, GN—which is the larger of the two
HO, OP—Our sixth planet has one moon, our fourth planet has 3, our first and third planets each have one.
GQ, QR—Our probe is in the orbit of your moon
ST—This updates the position of Arcturus shown in our maps.

Naturally, at present it would be impossible for astronomers to find out how many planets Epsilon Boötes actually has, but at least we do know that it is a double star as the message states. The complete details of Lunan's pioneering theory are beyond the scope of a book such as this.

'Although more evidence is required to support this hypothesis, the logic of Mr Lunan's work is of interest in its own right as a contribution to the problem of interstellar communication,' writes Kenneth W. Gatland, Editor of *Spaceflight* and Science Editor of the *Daily Telegraph* (London). On the whole, this exposé illustrates the probable existence of an alien space probe through which an attempt is being made to contact terrestrial humanity by returning our radio signals and conveying a graphic message.[6]

The French playwright Jean Giraudoux believed that thoughts fly between worlds: 'I know perfectly well that at this moment the whole universe is listening to us—that every word we say echoes to the remotest star.' (1945)

All evidence points to the certainty of man contacting other beings on unknown planets in space in the historically near future. As Arthur C. Clarke writes in his *Voices from the Sky*: 'The universe must be full of voices, calling from star to star in a myriad of tongues. One day we shall join that cosmic conversation.'

[6] *Spaceflight* (British Interplanetary Society), 4 April, 1973.

In the search for ETIs another possibility should not be overlooked. If galactic civilisations have at any time approached and visited our solar system, it would be logical for the crews of their spaceships to have built permanent bases on one or two of the outer planets or their satellites. The reason for selecting the fringe planets is obvious. After a spaceship has journeyed at fabulous velocities through galactic space, it would slow down at the periphery of our solar system, land on the first suitable planet and then begin the exploration of the other planets from this base. Using powerful ray machines they might later have bored through the geological strata of the Earth and constructed spacious underground bases in isolated mountain ranges, such as the Andes or the Himalayas.

13: *The Missing Planet*

A solar system could be described as a cosmic atom. Its nucleus would be the sun and its electrons, the planets. Most of the mass of the atom is contained in the nucleus. Likewise, almost the entire mass of the solar system is contained in the sun.

Owing to gravity and other factors the planets are spaced according to a law discovered by Bode and Titius. In the geometric progression 0, 3, 6, 12, 24, 48, 96, 192 each number is double the previous one, with the exception of the 0. Adding 4 to each result we obtain the computed orbits of the planets—4, 7, 10, 16, 28, 52, 100, 196. Actually, Mercury, Venus, Earth, Mars, Jupiter, Saturn, Uranus are placed at 3·9, 7·2, 10·0, 15·2, 52·0, 95·4, 191·8 which is very close to the figures of the calculated orbits under Bode–Titius law, only the orbits of Neptune and Pluto do not quite correspond to this rule. Surprisingly enough, there is no known planet in the orbit of 28.

Johann Bode, Director of the Berlin Observatory, was convinced that there should have been one. His concern for the missing planet had been shared for some time by Baron von Zach, an officer in the service of Duke Ernest II of Saxe-Gotha. The Duke encouraged this star-gazing hobby of his scientifically minded officer to the extent of building him an observatory. The baron was determined to find the missing planet. With military efficiency he convened an astronomical congress in Gotha in 1798, the first of its kind in the world. A permanent committee consisting of six German astronomers, with von Zach as secretary, was elected. They became known as the *celestial police* for their detective work in the heavens.

(*a*) Were outer fragments of a planet between Mars and Jupiter thrown into space during nuclear warfare to become asteroids now travelling on eccentric and irregular orbits?
(*b*) Did a chain reaction follow destroying that planet and transforming it into a ring of asteroids?

When their network of agents had been organised, on 1 January, 1801 news suddenly came from the Italian astronomer Piazzi that he had spotted an unidentified body.

Naturally, Bode and von Zach were elated. In order not to lose track of the tiny planet they decided to engage a mathematician capable of computing the orbit of the object. A student, Karl Gauss, who later became one of the greatest mathematicians of Europe, volunteered to solve the problem. He calculated the planetoid's position in the sky for the night of 31 December, 1801. On that date Ceres, a small planet 772 kilometres in diameter, was sighted in the sky. What is more, it revolved around the sun in the orbit predicted by Bode.

In 1802 Dr Heinrich Olbers, a member of this celestial detective team, discovered another little planet which they named Pallas, that had a diameter of 489 kilometres. The astronomical Scotland Yard was puzzled. They had expected to find only one missing planet of normal size but instead detected two undersized bodies. Dr Olbers suggested that a large planet might have exploded in the past, that the two planetoids were parts of this planet and probably many more fragments would be found floating in space.

Dr Olbers locked himself in his study in Bremen and began to calculate the possible trajectory of the debris of that unfortunate planet. Then one day he wrote to his colleagues, advising them to keep an eye on the constellations of the Virgin and the Whale. As a result of these recommendations, on 1 September, 1804 the astronomer Karl Harding of Lilienthal Observatory sighted Juno, 193 kilometres in diameter. Then in 1807 the doctor himself made his second discovery—Vesta, 386 kilometres in diameter. Although only half the size of Ceres, Vesta is the brightest of the asteroids, being occasionally visible to the naked eye. Today it is believed to be either a space iceberg or a little planet covered with ice.

Ever since that period the search for the remains of the dead planet has never ceased. We now know of the existence of some 60–70 asteroids of considerable dimensions, but none larger than 100 kilometres, except, of course, for those which owing to their greater size are known as the minor planets. The orbits of 2,000 asteroids have been calculated and catalogued. This wide asteroid belt contains thousands of rocks of different sizes, some of which are as small as a pebble or even a grain of sand.

Dr Olbers was convinced that the asteroid ring was the debris of a shattered planet. Even today some astronomers share his views whereas others subscribe to the theory that it is merely the material from which a new planet will eventually be formed. But why should the cementing of all this building material be a few thousand million years behind the formation of the other planets?

The theory of the agglomeration of the asteroid belt into a planet is further invalidated by the fact that due to collisions the rocks in the belt are constantly being smashed and broken up. The ring material is in the process of crumbling instead of fusing. Substance from the ring is also being lost in the form of meteorites falling on the Earth as well as on other bodies of the solar system.

Is it possible that our moon was once the mystery planet's satellite? A bombardment by a cloud of planetary fragments would account for the great number of craters on one side of the moon. An exploration of the lunar craters with a view to finding foreign material in them, could support or discredit this possibility.

In 1950 the lost planet was named *Phaethon* by Professor S. V. Orlov of the U.S.S.R. Academy of Sciences. Its greatest enigma is connected with its disintegration as an astronomical body. A planet does not break up unless it is destroyed by a collision with another body—an unlikely probability in view of the vast distances separating the planets. While the problem awaits a solution, the asteroid belt with its thousands of rocks, dust and ice is still spinning around the sun.

The larger minor planets are spherical due to the force of gravity which fuses the smaller components into a globe but the asteroids are simply enormous rocks of various sizes. Eros has the form of a long island measuring 24 by 6½ kilometres. Other asteroids, such as Icarus and Hermes, are huge mountains of stone floating in space. Although most of the planetoids and debris are concentrated in the zone between Mars and Jupiter, some asteroids do not follow this orbit. Hermes, Icarus, Eros, Hidalgo, Amor and Adonis which range from the size of a many-storied building to that of a city block, have eccentric orbits as well as those inclined to the ecliptic. Hermes touches the orbit of the Earth and then travels far beyond the asteroid belt towards the orbit of Jupiter. Icarus approaches the sun closer than does Mercury, becomes red-hot and then swings out beyond Mars towards its own rocky way. Eros reaches our orbit, flies towards the path of the asteroids, then turns back instead of going beyond. Hidalgo touches the path of Ceres and then ricochets towards Saturn in an elliptical and highly inclined orbit. Amor, Apollo and Adonis are 'earth grazers' as they come quite close to Earth during their revolutions around the sun. And the Trojans are two swarms of asteroids that had been captured by the strong pull of Jupiter. It, therefore, seems reasonable to conclude that the minor planets, the clouds of rock, dust and ice, as well as the asteroids with the eccentric orbits are all fragments of what was once a much larger body.

In the solar system there are streams of rock and dust that travel on irregular paths. Travelling in the same direction around the sun as the Earth, these particles are often pulled by the Earth's gravity and burn in the atmosphere at an altitude of 80 to 160 kilometres. Occasionally, larger 'shooting stars' break up from friction with the air as they fall at the speed of

70 kilometres per second and become meteorites—true visitors from space. Their arrival is unpredictable but welcome, allowing astronomers to examine them to determine of what stuff planets and comets are made.

On each 2–3 January one may expect shooting stars coming from the direction of Boötes and the Head of Draco, on 20–22 April another meteor shower arrives from a sector of the sky between Vega and Hercules, and on 20–22 August there are more meteors from the direction of the constellation of Perseus. The Taurids arrive on 8–10 November from the direction of Taurus, Auriga and Perseus. There are about 600 meteoric streams in the solar system, most of which are remains of disintegrated comets such as Biela, Tempel and Tuttle's comet. The material comprising the core of comets has been compared with 'dirty snow' or ice with grains of dust. Some comets melt passing too close to the sun, as a result all that remains is a cloud of meteoric dust.

But how can we be certain that the larger meteorites actually emanate from the asteroid belt? The Sikhote–Alin meteorite which fell in Eastern Siberia in 1947 provided reliable evidence that this is indeed their origin. It weighed originally about 1,000 tons before breaking into millions of fragments. Eyewitnesses described it as a ball of light as brilliant as the sun. A map of the craters and holes made by these meteorites as well as the trajectory of their flight in the sky, enabled Professor V. G. Fesenkov of the U.S.S.R. Academy of Sciences to trace their source directly to the asteroid belt. Photographs of the trajectory of the Pribram meteorite, which fell in Czechoslovakia in 1959, made it possible to calculate that it too came from the asteroidal zone.

If meteorites came from interstellar space, they would have described parabolic or hyperbolic orbits. Yet they are known to travel in closed elliptical arcs which proves that meteorites belong to this solar system, particularly to its asteroid ring between Mars and Jupiter. Consequently, fragments from the shattered fifth planet of the solar system are available for study.

There are three basic types of meteorites, nickel–iron, stony iron and stone. When analysed, iron meteorites show that in density they can be compared only with the material of the Earth's core which also consists of nickel–iron. This has been proven by the presence of *Widmanstätten figures* on the surface of these meteorites, which suggests they have been submitted to extremely high pressures and temperatures. This contention is also supported by the occasional discovery of diamonds and graphite in meteorites, both of which require the activity of tremendous pressure during their formation.

The chemical composition of stony meteorites generally resembles that of terrestrial rocks. However, there do exist slight yet important differences. Minerals such as iron sulphide, phosphorus in a complex compound with iron, nickel and cobalt and several others are indigenous only to meteorites. Some stone meteorites incorporate radial assemblies of crystals known as chondres. These chondrites are not found in earth materials although they do look very terrestrial. Stony meteorites may also contain microscopic bubbles of carbon dioxide and water in their microstructure.

The largest meteorites that reach the Earth belong to the nickel–iron type. The Hoba meteorite, found in Southwest Africa in 1920, weighs 70 tons. The Ahnighito meteorite which fell in Greenland weighs $36\frac{1}{2}$ tons. No stony meteorite weighing more than a ton has ever been discovered, probably because they are either shattered in space in collisions or break up passing through the Earth's atmosphere. The Norton County meteorite which fell in 1948 is the largest in this category.*

There is another type of celestial refuse, the dumping of which no one has ever seen in known history. It is named

* Norton County, Kansas, U.S.A.

tektite which is derived from the Greek word *molten* owing to its resemblance to molten drops of glass. They can easily be mistaken for bits of bottle glass. Some are amber in colour, others green or dark brown. Chemically, they are composed of silica with oxides of aluminium, iron, magnesium and calcium. The shapes of tektites would suggest that they melted during their flight through the atmosphere and fused on impact.

Tektites differ from meteorites and it is not certain that they originate in the asteroid belt. Firstly, they are composed of a special glass that has a high melting point; secondly, they are found only in specific places on Earth and nowhere else. Tektites exist in Australia, the Philippines, South East Asia, Libya, Czechoslavakia, the Ivory Coast, Texas and Georgia. The Australites have been recognised to be the youngest tektites—they fell only 5,000 years ago. The Bediasites from Texas are the most ancient, having arrived about 45 million years ago. No tektites have fallen during the last 5,000 years, nor are there any tektites in geological strata older than 45 million years.

It appears that the asteroid belt is not the source of tektites. Measurements on cosmogenic isotopes in tektites indicate that they spent only a short time in space. It is thought that they could not have originated further away than the lunar orbit.[1] Spherical glassy bodies have been found on the moon but they are considered to be the product of the intense heat caused by meteorite impact. Brian Mason in his article *The Lunar Rocks* has demolished the hypothesis of terrestrial tektites' origin on the moon.[2] Thus though the problem of tektites is still unsolved, at least it is now known that they do not come from the asteroid belt as do most meteorites.

Meteorites have been detected in 50-million-year-old geological layers. Collisions of asteroids with our planet must

[1] British Museum, Meteorics Room, Exhibit 10, South Kensington, London.

[2] *Scientific American*, October, 1971.

have occurred during earlier epochs which would explain the existence of huge craters, such as the Brent Crater in Canada, 500 million years old. But to the best of our knowledge, none of them left any hardware.

The largest remnants of Phaethon—Ceres, Pallas, Vesta and Juno—would cover France, Belgium, the Netherlands and Western Germany. However, if the asteroid ring with these planetoids and the Trojans in the Jupiter orbit were all welded together, the result would produce only a very small planet. Nevertheless, this does not rule out the possibility that Phaethon may have originally been a large planet in view of the mass of material lost in space. The quantity of the missing planet's rocks which have fallen on Earth, Mars or other planets can hardly be estimated. We know, at least, that an average of 3 tons of meteoric material falls on the Earth each day.[3] If the rate of fallen meteorites were the same in the past, over one million tons of meteoric matter would, therefore, have fallen on the terrestrial surface alone in the course of the last thousand years—and most of it would have come from the zone of the asteroids.

One wonders whether the story of this tragic planet may indeed be coded in the meteorites that come from the asteroid belt. First of all, let us bisect our own planet. The Earth possesses a dense nickel–iron core about 6,800 kilometres in diameter, enveloped by a 2,900-kilometre layer of molten rock which is denser near the nucleus and lighter further away from it. The planet is covered by a thin crust of 50 kilometres. The major bulk of terrestrial material comprises rock and nickel–iron, whilst the crust constitutes only a small portion of the substance of the Earth. The outer planets have a different structure consisting of a rocky core and thick layers of ice and gas. The spectral analysis of meteors flashing in the sky discloses their chemical components. There are two types of

[3] *Nauka i Zhizn* (Science and Life), No. 10, 1972 (U.S.S.R.).

composition—one that shows strong lines of calcium, chromium, magnesium, manganese, and the other iron and nickel. Type One is identical to the composition of stony meteorites and Type Two to that of iron meteorites. Nickel–iron spectra are encountered in only 25 per cent of analyses, indicating that there are four times as many stony meteoric particles in space as there are nickel–iron fragments. In the Earth, rocky material is also more abundant than nickel–iron in approximately the same ratio. Apparently Phaethon consisted of a metal core and a rocky mantle very much like the Earth. This similarity of the components of the Earth with meteorites is recognised by astronomers. According to A. A. Moss,

> If, therefore, the Earth were to break up into fragments, and we were in the hypothetical position of being able to examine these fragments, we should expect to find pieces of nickel–iron, pieces of rock, and from the region of the discontinuity of 2,900 kilometres pieces that consist of mixtures of nickel–iron and rock (stone). This is precisely what we encounter in meteorites and it is customary to divide meteorites broadly into three classes—irons (or siderites), stony-irons (or siderolites) and stones (or aerolites).[4]

The average density of our planet is about 5·5 times that of water. It is extremely high in its centre—10 to 12 times that of water and varying between 2·5 and 3·5 for surface rocks. Stony meteorites have almost the same density as the rocks of Earth, while the density of nickel–iron meteorites is also surprisingly high. On the basis of this, the Soviet meteorics expert Fedinsky arrived at the following conclusion: 'The density of iron meteorites may be compared only with the material that comprises the central core of the Earth.'[5] We are irresistibly

[4] A. A. Moss, *Meteorites*, British Museum, 1971.
[5] V. Fedynsky, *Meteors*, Moscow, 1959.

drawn to the argument that a ferrous meteorite is a metallic fragment of the kernel of the perished planet.

Stony meteorites provide many surprises. Among them is a type known as the carbonaceous chondrite, which contains carbon and bound water. In 1961 the American scientists Drs Bartholomew Nagy, Warren G. Meinschein and Douglas J. Hennessy conducted a laboratory analysis of small fragments of the Orgueil meteorite which fell in France in 1864 and belonged to this chondrite category. Since carbonaceous chrondrites are extremely rare, the analysis was performed with great caution and thoroughness. It showed that 15 per cent of that chondrite consisted of paraffinoid hydrocarbon molecules of organic type. When the puzzling material was isolated, microphotographs revealed the presence of fossilised one-cell organisms. They are known as dinoflagellates or chrysomonads which inhabit the waters of lakes and seas.

Drs B. Nagy and G. Claus were the first to establish the presence of these organic microstructures in carbonaceous meteorites. In addition to the Orgeuil specimen, other chondrites of a similar type from Central Africa (Ivuna), India (Tonk) and France (Alès) have been analysed. These sensational findings were independently confirmed by Dr Frank Staplin of Canada, Professors Erdtman and Skuja of Sweden, Dr Robert Ross of England, Professor A. Papp of Austria, Dr Cholnoky of South Africa and Dr B. V. Timofeyev of U.S.S.R. They have proven that some chondrites contain hydrocarbons, fatty acids and even amino-acids.

Certain tests provided photographs of cell fossils of a form unknown on Earth. But in a piece of the Orgeuil meteorite Dr R. Ross photographed two mushroom-shaped microscopic bodies like the micro-fossils in pre-Cambrian rocks. Naturally, some scientists refused to accept these fantastic disclosures made by their colleagues. Professor Nagy argued: ‘If you conduct microbiological tests on these chondrites yourselves, you will arrive at the same results.' And many did. Professor Melvin Calvin, the first man ever to attempt to re-create the genesis of life in a laboratory, examined fragments of a meteorite which had fallen in Kentucky in 1950. His analysis showed the presence of molecules of the aromatic heterocyclic type resembling the pyrimidines and purines present in genetic material. He commented: 'The samples indicate that the same evolutionary processes that unfold on Earth have gone on somewhere else.'

The last word in this controversy was uttered by the Ames Research Centre of NASA, whose Director is the noted biologist Dr Cyril A. Ponnamperuma. In search of organic life the team was requested to perform a special analysis on an Australian chondrite which had fallen near Murchison, Victoria, in 1969.

It must be emphasised here that the living world is composed of asymmetrical molecules. This could be compared to mankind—most people are right-handed yet some are left-handed. The left-handed amino-acids predominate on Earth whereas the right-handed type of polymers is relatively rare. The distribution of amino-acids in the Murchison meteorite was totally unearthlike with the right- and left-sided forms appearing in equal quantities. Contamination by terrestrial amino-acids would have added the laevorotatory but not the dextrorotatory type, which is scarce on Earth.

On this crucial issue Professor Ponnamperuma wrote: 'Further, it was apparent that there were a number of amino-acids which are not commonly found in proteins. These too could not be ascribed to terrestrial sources, and had, therefore, to be indigenous to the meteorites.' And he added another important detail: 'The meteorite appeared to be enriched with a heavy isotope of carbon, further confirming its cosmic credentials.'[6]

[6] C. Ponnamperuma, *The Origins of Life*, London, 1972.

The scientist stated that the Murchison meteorite was one of the pieces of an exploded planet and that it constituted 'the first conclusive proof of extraterrestrial chemical evolution—the chemical processes that preceded the origin of life'.[7]

The late astronomer Dr Harlow Shapley wrote that: 'If they (the asteroids, A.T.) did originate, as many believe, from the bursting of one or more planets, we may rightly surmise that the organic compounds recently discovered in analyses of meteorites point to the former existence of a planetary site of life, a site that once existed on the outer fringe of the sun's liquid water belt.'[8]

Did the sun radiate more heat when it was younger in stellar sequence and did it provide Phaethon with enough warmth in its orbit beyond Mars? Such a theory would similarly explain the global tropical climate that the Earth enjoyed in prehistoric eras.

When did the planet Phaethon explode? Estimates range from 50 to 500 million years, based on the calculation of the earliest falls of meteorites. The oldest meteorites found in geological strata are dated to a period 50 million years ago, while the Brent Meteoric Crater in Canada is considered to be 500 million years old. Edward Anders of the Enrico Fermi Institute for Nuclear Studies discovered that many stone meteorites had been storing gaseous products of radioactivity (helium 4 and argon 40) for only 400 million years. At the moment of the explosion of Phaethon heat would have liberated the helium and the argon which was stored up until that time.

A Soviet astronomer, Dr Felix Siegel, has proposed a curious theory regarding the end of Phaethon. He believes that the destruction of the planet may have been due to artificial rather than to natural causes. The astronomer speculates that on the planet existed an advanced race of rational beings who

[7] *Christian Science Monitor* (Eric Burgess), 4 December, 1970.

[8] H. Shapley, *The View from a Distant Star*, New York, 1963.

eventually achieved a technological civilisation. They discovered nuclear energy and constructed a stockpile of hydrogen bombs. In a fratricidal war a mass explosion of hydrogen bombs created a chain reaction, igniting the hydrogen of the seas, thus causing the entire planet to explode.[9]

A Russian writer, Alexander Kazantsev, exploited this theme for his science fiction novel *The Phaets*. The visiting physicist Niels Bohr was asked by Kazantsev in Moscow in 1961: 'If a hydrogen bomb is exploded in the depths of an ocean, and hydrogen is instantly converted into helium, could this bring about the explosion of the entire ocean?' Bohr thought awhile and then slowly said: 'Such a catastrophe is possible . . . although some scientists deny the possibility.'

Dr F. Siegel has referred to the statement of Professor A. P. Vinogradov, who declared at the International Chemical Congress in 1965 that chondrules—the tiny beads of magnesium iron silicates found in stony meteorites—could be produced experimentally in underground nuclear tests.

According to Siegel's theory, nickel–iron meteorites would be the remains of the core of Phaethon, stony-irons would come from the border zone where the rock mantle touches the iron core, and stony meteorites—from the mantle itself. Carbonaceous chondrites, the lightest of all meteorites, would originate from the crust with its soil and traces of life.

Are Hidalgo or Hermes, which travel in eccentric orbits far from the asteroid belt, chunks of the crust from the first target site in 'enemy territory'? Was there a partial detonation followed later by a chain reaction of the whole planet which exploded radially and uniformly? If so, the asteroids on eccentric orbits should be mainly composed of rock and soil emanating from the initial explosion in the area attacked. The entire site would have shot into space like a launched rocket on a pad—this being the portion of the planet that was still intact.

[9] F. Siegel, *Zhizn v Kosmose* (Life in the Cosmos), Minsk, 1966.

After a brief period and more partial detonations, finally came the chain reaction which spelled doomsday for Phaethon.

It is maintained by some astronomers that if a body orbiting the sun had exploded, the orbits of its fragments should intersect somewhere at the point of the past cataclysm. Although the paths of the larger asteroids, Ceres, Pallas and Juno, fulfil this requirement, thousands of asteroids do not. However, if one considers the perturbations caused by the planets, particularly Jupiter, during the course of millions of years, it is small wonder that most of the asteroids have non-intersecting orbits.

A space probe could check the material of asteroids with long elliptical orbits to ascertain whether they do consist mostly of rock, while an exploration of Mars could establish the presence of meteorites on its surface and compare them with those on Earth—all the more significant since Phaethon and Mars were neighbours. It is not impossible that some of the moons of Jupiter and Saturn might be captured remnants of the exploded planet.

Lunar soil has presented a riddle to astronomers. The surface of the moon is covered by carbonaceous chondrite dust. This residue constitutes 2 per cent of lunar soil and is totally dissimilar to the rocks on the moon.[10] Did the chondrite dust from Phaethon settle on the moon when it was much closer to it—probably as an independent planet?

The nuclear chain reaction of a planet is not simply an alarming subject to be expounded in a science fiction thriller. It has actually occurred on Earth on a small scale! Dr Francis Perrin, former chairman of the French High Commission for Atomic Energy, has recently submitted to the French Academy of Sciences two papers which outline a chain reaction that took place in Africa many millions of years ago.[11] He described uranium from the Oklo mine in Gabon, Africa, stating that it had

[10] Walter Sullivan, *New York Times*, 7 January, 1972.

[11] Walter Sullivan, *New York Times*, 26 September, 1972.

a peculiar composition, depleted in the uranium 235 as if it had been burned in nuclear chain reaction. This was confirmed by the presence of four rare elements—neodymium, samarium, europium and cerium—in forms that comprise the typical residue of uranium breakdown, after fission.

These chain reactions began at least several hundred million years ago and continued for several million years until the fissionable uranium was burnt out. Nuclear scientists have not come to any agreement as to what had triggered the chain reactions in the Earth's crust. But they were certain that this natural atomic pile had been subjected to a pulsating chain reaction. Fortunately for Africa and perhaps for the planet Earth, this was only uranium. A hydrogen chain reaction would have spelt doom. That is what probably happened to the world of Phaethon.

Few astronomers would subscribe to the idea of a nuclear chain reaction having destroyed Phaethon, yet most of them seriously accept the possibility that a planet in our solar system between Jupiter and Mars did actually disintegrate due to some unknown cause, and that its debris formed the asteroid belt. Fifty years ago Professor Kiyotsugu Hirayama of the Tokyo Imperial University wrote that 'the asteroids, so small and numerous, may be fragments of larger bodies which existed originally'. He was quite certain that this cloud of meteoric matter had formed from 'the breaking up of a single asteroid'.[12]

Most of today's astronomers agree with Dr John A. Wood's conclusion that 'the evidence in favour of parent meteorite planets is very strong'.[13] However, there is no unanimity of opinion as to the size of the disrupted body or bodies.

The dossier of the Missing Planet still remains open. But if Dr Olbers of Bremen could examine the file of Phaethon today,

[12] *Annales de l'Observatoire de Tokyo*, Appendix 11, January, 1923.
[13] J. A. Wood, *Meteorites and the Origin of Planets*, New York, 1968.

he would be surprised to discover just how much data has been accumulated since Napoleonic times.

While the question of the origin of the asteroid ring awaits a solution by the academic world, the presence of organic molecules on meteorites and the ratio between iron and stone meteorites have also to be explained. Appalling as it is to envisage the atomic suicide of a past planetary civilisation, it is not devoid of significance and deserves reflection in this strife-torn world of today.

14: *Cosmic Civilisers*

Eventually our spaceships will reach the outer planets of the solar system, build bases on them, and then venture further into interstellar space in photon rockets.

On some unknown planet warmed and lighted by a white or yellow sun our astronauts may discover primitive life. Should primates exist on that planet the astronauts would separate them, concentrating their attention on the most intelligent. How to use a tool, make a fire, heal wounds, to count—these would be some of the first lessons the earthmen would give. The starships would come again and again, and the men from Earth would then impart more advanced instruction, such as how to identify the stars or to convert ideas into signs and sounds. During one of these missions an astronaut might even be ordered to couple with an ugly, hairy animal-woman, so that the evolution of that primitive race may be accelerated by artificial methods, and the species thus improved.

These half-men would then venerate the astronauts as gods, which would be a natural reaction in such circumstances. Then, after completion of their assignment, the explorers would depart for their home planet, leaving the planetary menagerie to work out its own destiny in the garden of evolution. During decades, centuries and thousands of years, the inhabitants of the distant planet would repeat stories of visitations of superior messengers from the stars, who had bestowed upon them the blessings of culture.

Since this programme would have been carried out over the entire planet, similar legends would circulate in different parts of that world. All these verbal traditions would then be committed to writing and thus become a part of the cultural heritage. After long ages, the wise men of this reasonably advanced civilisation would begin to express their scepticism towards all those myths and holy scriptures depicting messengers from the stars. The distinguished descendants of those clever apes would then maintain that they, of course, could fly in the sky and perform half the feats that the 'gods' or 'angels' were reputed to have performed, but that the legends themselves were only flights of fancy.

There is evidence to suggest that all this actually took place on Earth in a prehistoric era. The history of those space missions from Planet X would therefore be written in mythology, folklore and religion. Here we must ask a question. Can myth be really taken for a fossil of history? Only to a certain extent. With the passage of time verbal tradition becomes more fanciful. And there are many myths which have no connection with history, such as those that are interpretations of the origin of the world. The ancient Greeks used to say that myth was a truth stated in a metaphor, and in spite of being the authors of a voluminous mythology, always remained the most rational of peoples.

'Myth is history in disguise,' said Euphemerus in the fourth century before our era. Aristotle wrote: 'Our forefathers in the most remote ages have handed down to their posterity a tradition in the form of a myth.' In the nineteenth century no European scholars believed that there had ever been a city named Troy as described in Greek legends. When Schliemann discovered the ruins of many Troys, one built over another, the historians were forced to admit their error. In the light of this example, it is fairly obvious that behind a myth there is often an actual historical event. History, mythology and the sacred scriptures of most peoples contain a wealth of material portraying the descent of gods to Earth and their life among men.

The Dogon of Mali in Africa worship a pyramid with steps

leading to a square platform on top where, according to one of their legends, the sky gods landed on each of their visits, long, long ago. One of these beings showed them how to divide up their land and taught them how to cultivate it. The Dogon priests speak of an epoch when the gods came regularly 'to play on Earth'. Their ancient tradition speaks of the 'dark brother' of the star Sirius.[1] This is most extraordinary as Sirius does have a companion star which, however, is visible only through the most powerful telescopes. It is also a mystery why certain cave carvings in the Mediterranean basin depict the Pleiades as ten stars, whereas only six or seven are visible on a clear night, with excellent vision. It takes a telescope, however, to see ten or more stars in that constellation.

The astronomer-priests of Babylon had stepped pyramids, the pinnacles of which were reserved for the sky-beings coming down to Earth. The pyramids of Chichen Itza and Tikal in Central America are very much like those of the Dogon and of Babylon. Again, their purpose was similar—to provide specific sites upon which the celestial visitors could land. Considering the isolation of the Old World from the Americas for thousands of years, it is a wonder that such identical structures and legends should have originated independently.

The Nuba people of Sudan hand down a myth according to which in the beginning the sky was so near that man could actually touch it. They speak of a rope that was suspended from the sky by which certain people could ascend to and descend from heaven. According to Geoffrey Parrinder the Shilluk of the Sudan believe 'that in the olden days men used to be able to get up to the moon by a road, but they became too heavy and could no longer use it'. It seems that modern man has recently reopened this highway to the moon.

The priests of ancient Egypt alluded to the *First Time* when gods actually lived on Earth during the Golden Age. The

[1] J. Servier, *Je ne crois pas au progrès*, Planète (magazine), No. 18, 1964.

Egyptians erected obelisks dedicated to the Sun-god Re which had gilded tops called Benben. Were these obelisks inspired by space rockets in which the sky gods arrived?

The *Book of the Dead*, the 3,500-year-old Egyptian bible, contains strange words. One passage speaks of 'those who with their knowledge reach the vault of the sky'. Another chapter even mentions 'those who live among the stars'. Thoth was the great culture-bearer of Egypt who gave the people of the Nile the first elements of science, mathematics, literature, history and medicine. After completing his task he returned to the starry heavens.

On the other side of the Atlantic the sun cult was as prevalent as it was in Egypt. According to Garcilaso de la Vega, God the Sun sent one of his sons and one of his daughters—Manco Copac and Mama Ocllo—to transform animal-men into rational beings. These heavenly messengers taught early mankind how to build cities, raise food plants and 'live as men in reason and comity'. The city of Cuzco, with the most monumental stonework in the world, still stands as a mute witness to the origin of this legend. We should be grateful to both the pre-Incas and the Incas for having been the first to cultivate half the vegetables we eat today.

In Mexico, the god Tezcatlipoca decided to give mankind the gift of music and commanded his messengers to go to the High House of the Sun:

> Go, bring back to Earth a cluster, the most flowering,
> Of those musicians and singers.

In Honduras there is a legend that tells of the White Woman of unparalleled beauty who came down from heaven to the town of Cealcoquin. She commanded the natives to build her a palace ornamented with 'strange figures of men and animals', and to deposit in the temple a stone inscribed with unknown characters. This celestial virgin gave birth to three sons who

(*a*) Arrival of Quetzalcoatl, the culture hero of Central America, in a serpent-like craft. (A mural in the Palacio Nacional in Mexico City.)
(*b*) Ancient Egyptian painting of a celestial messenger coming to Earth.

became princes of the kingdom. Many years later the queen returned to her heavenly home.

Quetzalcoatl, the Plumed Serpent, disembarked from a winged ship and brought to Mexico the benefits of civilisation. He imparted to the Indians the knowledge of astronomy and mathematics, and they devised with him the most precise calendar in existence, even today. He must have also taught them the principles of architecture as the world's most massive pyramid, so far as its base is concerned, was constructed by the Mayas. Due to lack of co-operation, the messenger departed to the sky in a spectacular exhibition of fiery flashes. In the Palacio Nacional in Mexico City one may still see murals which depict Quetzalcoatl flying in a serpent-like craft. Mexico has not forgotten her debt to Quetzalcoatl.

The memory of an era of celestial guardians is deeply ingrained in the consciousness of India. Even today the Hindus point to another half-forgotten Space Age when the gods unexpectedly arrived from the stars. That is why it is the sacred duty of every Indian to treat a guest as if he or she were a messenger from heaven. In fact, this custom is typical of most of Asia for the same reason. Hospitality is also highly recommended by the New Testament: '. . . there are some who by so doing have entertained angels without knowing it' (Heb. 13:2). Ovid relates how Jupiter and Mercury took on the appearance of poor wayfarers and paid a visit to Baucis and Philemon of Phrygia. Just before their departure they declared: 'You have been hosts to gods.'

The Bible represents angels as having existed before man. They are called the *upper ones* or the inhabitants of heaven. If angels were denizens of other worlds older than ours, then the biblical statement might be perfectly correct. In the Old Testament it is stated that the giant *sons of God* coupled with the *daughters of men* who produced *mighty men.* These activities may have been nothing other than genetic experiments. Lot entertained two angels in his house in Sodom with matzoth and *they did eat* (Genesis, XIX. 3)—it is just this precise detail that discounts the supposition that those particular angels were not physical beings.

The Hebrew Kabala refers to Metatron, the leader of the *yorde merkabah*, celestial voyagers, who sometimes descended to Earth in their chariots, and who, according to the Kabala, knew all the secrets of the starry heaven. Furthermore, Rabbi Simon Ben Jachai stated in the *Book of Zohar* that the Masters of Kabala had always cherished a belief in the unpredictable arrivals of superior beings from the sky.

Since the Kabala consists largely of Jewish gnosticism, it is therefore interesting to read *The Vision of Aridaeus* by Plutarch of Chaeroneia, also a gnostic, where he describes a swift voyage in space, travelling as smoothly 'as a ship in calm weather'.

'Aridaeus saw naught save the stars, they were, however, of a stupendous size and at enormous distances from one another, and poured forth a marvellous radiance of colour and sound.' It would not be wise to reject this 2,000-year-old record in view of the scientifically correct information it contains regarding the size of the stars, their distances from each other, and the emissions on different frequencies.

Interesting tales are included in the *Book of Enoch*[2] regarding this two-way traffic in space. Extracts from this old book, written in the pre-Christian times, read almost like a science fiction story. At first the patriarch met two tall men—'such as I have never seen on earth', who told him that—'today thou shalt ascend with us into heaven' (2, I). Realising the serious character of this coming space voyager, Enoch summoned his family to say 'farewell': 'Hear me, my children, for I do not know whither I am going, or what awaits me' (2, II).

Then the two strange men took Enoch in a celestial chariot to a very high altitude. Flying above the clouds, Enoch's emotional reaction was quite natural: 'And lo! the clouds moved. And again (going) higher I saw the air and (going still) higher I saw the ether, and they placed me in the first heaven' (2, III). The words 'air', 'ether' and 'first heaven' may perhaps be translated as 'atmosphere', 'space' and 'first planet'. Then the

[2] W. R. Morfill and R. H. Charles, *The Book of the Secrets of Enoch* Slavonic Enoch), Oxford, 1896.

next paragraph does make sense: 'And they brought before my face the elders and the rulers of the stars' (2, IV). Evidently, these planetary leaders possessed great wisdom, as according to Enoch, 'these orders arrange and study the revolutions of the stars . . . and they arrange teachings and instructions' (2, XIX).[3]

This science must have been too advanced for the earth-people of that epoch: 'Wisdom went forth to make her dwelling among the children of men, and found no dwelling-place. Wisdom returned to her place and took her seat among the angels' (3, XLII).

The *Mahabharata* contains verses so amazingly in accordance with modern astronomy that one wonders where the Brahmins obtained their knowledge:

> Infinite is space where the perfect ones and gods dwell,
> It is full of their delightful abodes without number.
> Above and below their course the sun and the moon can not be seen,
> The resplendent gods, the fiery-coloured ones, shine there like suns,
> But even they fail to see the edge of the sky so mightily spread.
> So infinite, boundless and difficult it is to reach.
> The resplendent, shiny ones conquer this space further and further,
> Though it is immeasurable even for the gods.[4]

These lines from the epic of India show that galactic communities have been engaged in space travel since time immemorial. The same Sanscrit source also contains a description of an archaic airship: 'A wondrous crystal chariot able to travel in the air, of the kind that the gods use in airy space.' It is written

[3] R. H. Charles, *The Book of Enoch* (Ethiopic Enoch), London, 1917.

[4] Translation by Prof. B. L. Smirnov, Turkmen Academy of Sciences (U.S.S.R.).

that this aerial car was able to ascend to *Naksatra Mandala*, in other words—the stellar spaces.

Garuda, the man-bird of Vishnu, is alleged to have explored not only the moon but even the Pole Star. The memory of this space pioneer is kept alive in the Far East even today—Indonesia's national airline is named after him—Garuda Airlines. One of the Vedic hymns, dedicated to Varuna, intimates that there had even existed some sort of galactic police in space:

> Whoever far beyond the sky should think his way to wing,
> He could not there elude the grasp of Varuna, the king,
> His agents descending from the skies glide all this world around,
> Their thousand eyes all-scanning sweep to Earth's remotest bound.

Reading this verse, it would seem that the ancients were aware of the same sort of reconnaissance activities as those of today's controversial UFOs.

China used to be known as the Celestial Empire and no emperor could rule unless he received the *Mandate of Heaven*. In fact, all ancient dynasties reigned by the authority of *Dieu et mon droit*. Whether in China, Egypt or Peru, the formula was always the same—first ruled the gods themselves, who were then followed by their offspring from mortal women in the so-called solar or celestial dynasties.

Much of African mythology mentions a god who lived on Earth helping man. According to these myths, the sky benefactor, angered by the evil ways of men, departs heavenward. The curious thing about these legends is their striking similarity to the myths of Re, Zeus and Quetzalcoatl.

As Dr Raynor C. Johnson writes: 'We may speculate whether the Divine Society has wholly evolved within our world-system, or whether, since we know there are world-systems older than ours, it may not have had from its earliest beginnings the guidance and direction of advanced beings from other systems.'[5]

Greek mythology is rich with tales of what may be termed *interplanetary adultery*. Zeus, the king of the gods, was so much venerated by the Greeks that he could get away with anything. His first terrestrial concubine was Niobe, daughter of Phoroneus. Their child Argos founded a city named for him. Zeus was then involved in a scandalous affair, the rape of Europa, after which she bore him three illegitimate sons. Leda, wife of Tyndareus, unfaithful to her husband, became the mother of Pollux and Helen by Zeus. Then there was Danae, who bore Zeus a son named Perseus. Perhaps the amours of Zeus were part of a well-planned genetic programme. He had to seduce Amphitryon's wife Alcmene 'to produce a son who would one day be a powerful protector of gods and men alike'. This was Hercules. Could this interplanetary intercourse have had as a purpose the establishment of the nucleus for the future European civilisation?

Apollo, Hermes and Ares also had affairs with mortal women who bore them children. Aphrodite, the goddess who travelled in a golden chariot to the celestial palace of Zeus—the planet Jupiter, tempted Anchises, the Trojan, and bore a child who was known as Aeneas, the pious.

The people of the Nile paralleled the Greeks in fabricating these amazingly strange fables of gods marrying mortal women. It is an historical fact that within the enclosure walls of all the great temples of Egypt stood *birth houses*. Traditionally, they were used by the gods whenever it was necessary to establish the divine descent of a future pharaoh.

It appears that the deeds of the stellar torchbearers of galactic culture are also described in some rare books of Asia. When L. Austine Waddell visited a monastery at Gyantse, Tibet, in

[5] R. C. Johnson, *Nurslings of Immortality*, London, 1957.

1903 he was shown a huge library of known Buddhist writings and also works unknown even to the monks. In 1966 I visited the Ghum Monastery near Darjeeling in the Himalayas. Its main attraction is a colossal Buddha whose forehead is encrusted with an enormous diamond the size of an egg. But more intriguing were the hundreds of scrolls and manuscripts of great antiquity, stored in pigeon-hole shelves. Of the history and contents of some of these Tibetan scrolls the lamas knew nothing. I also learned that certain valuable books had been removed during the early twenties by the head lama of this monastery to an isolated retreat in Tibet 'in order to preserve the gems of age-old wisdom'. Following this personal experience I could accept Helena Blavatsky's discovery of the mysterious *Book of Dzyan* in south Tibet about 100 years ago.

Citations from this archaic manuscript which is written in Senzar or early Sanskrit, portray an expedition of cosmic civilisers to Earth:

> Said the Earth: Lord of the Shining Face, my house is empty, send thy sons to people this world. (Stanza 1 : 2)
>
> Man remained an empty senseless being. (Stanza 4 : 17)
>
> Dark grew the space between globes. Radiant the two worlds became. (Series II, Stanza 5)
>
> Finding the distance just, flashed like a sheet of intermittent flame. The Watchers began their task. (II. Stanza 5)
>
> The serpents who redescended, who taught and instructed. (Stanza 12 : 49)
>
> The fifth race was ruled over by the first divine kings. (Stanza 1 : 48)

These selections give the picture of the arrival of cosmic culture-bearers from another planet in order to accelerate the development of primitive man. The description of their flight in space when the two planets became luminous, is most convincing. The astronauts flew in a serpent-like vehicle at tremendous velocity. Upon landing on Earth they founded the first divine dynasties.

The plot follows the same general pattern characteristic of all ancient legends in that the evil ways of man discourage the gods to the abandonment of their mission. The *Tongshaktchi Sangye Songa* of Northern Buddhists records: 'The kings of light departed in wrath. The sins of men have become so black that Earth quivers in her great agony.'

The Brahmin writings of India allege that divine spacemen had performed biological experimentation with apes. Hanuman, the monkey god of the *Ramayana,* who is still revered in India, was conceived when Shiva approached Anjana, the ape, and gave her a sacred cake. Could this have been some sort of genetic pill? At any rate, some time later Hanuman was born, a strong and intelligent super-monkey. He was not the only guinea pig in this hominisation programme of the celestial beings as he was followed by other simian heroes—Sugriva, Brahaspati, Bali, Tara, Gandha, Madana, Nala, Mila and Sushena.

Neither should genetic tests with bears be overlooked as Vishnu's offspring was called Jambavan, the king of the bears, Hindus believe that Hanuman still lives in inaccessible mountain regions and, considering the strange resemblance of the yeti to a bear or to a giant ape, one wonders whether it is a surviving remnant of these breeding schemes of the cosmic civilisers.

One of the latest reports on the Himalayan yeti comes from the American zoologist Jefferey McNeely who, on an expedition in Nepal, took casts of footprints which he compared to the *tracks of a primate* and which measured 22 by 12 centimetres with a wide, rounded heel. The explorer remarked that the yeti walked right next to one of his expedition's tents without disturbing its occupants.[6]

Many scientists deny the possiblity of crossing a species

[6] Reuters, Katmandu, Nepal, 5 January, 1973.

from one planet with a species from a different planet, and they are right within the present knowledge of modern biology. There was a time when the giraffe was believed to be a progeny of the camel and the leopard but these fanciful notions have long been discarded. On the other hand, artificial hybridisation has been successful in the production of new fruits. For the changing of the structure of animal organisms all that is lacking is time—thousands, even millions of years. However, mutants and hybrids have already been produced on the cellular level.

'Today we must seriously study all information relative to the experimentation with hybrids of man and mouse, man and rat, and even man and chicken,' writes U. Kolesnikov.[7] These cultures of somatic cells are now brewing in sealed containers in a number of biological laboratories throughout the world. No matter how weird this disclosure may seem, it supports the hypothesis that a highly competent team of biologists from another civilisation in space could have produced *homo sapiens* from the primates. Most decidedly, they might have needed millions of years to achieve this hybridisation—first in order to improve the quality of the original primate, and then to cross with an advanced form of animal-men themselves, creating mutations, and waiting and watching during untold ages.

A number of different genetic projects could have been initiated by the galactic astronauts and scientists. According to the myths of Indonesia, the Marqueses, Hawaii and Tahiti the first men on Earth were given birth to by a celestial couple. In the Caroline archipelago the natives insist that primordial man multiplied by budding as plants do. On Easter Island an old legend describes the first human being as coming out of an egg. This practice of eliminating sex in breeding by such children of Nature is difficult to understand.

Modern palaeontology draws its deductions on the appearance of man on the basis of unearthed bones and artifacts. This

[7] *Nauka i Zhizn* (Science and Life), No. 3, 1972 (U.S.S.R.).

science was even more full of puzzles until the discovery of the Leakey skull. The prehistoric ancestors of man must have been extremely scarce as in only nine sites have human bones been located, covering a period from 2,500,000 to 300,000 years. But more and more skeletons and artefacts emerge from around 60,000 years ago. There seems to be considerable evidence that 40 centuries ago modern man appeared on the scene in a completely finished evolutionary form. Was some sort of hybridisation carried out by visiting cosmic scientists around 50,000 years ago?

There may be much truth behind the visionary film *2001—A Space Odyssey* by Kubrick and Clarke, in which a superior galactic intelligence gives an evolutionary impetus to the apes of Earth.

Hundreds of myths could be added to the mass of evidence illustrating a possible visitation to Earth by astronauts from an advanced stellar system on their mission of exploration and scientific aid. Admittedly, none of these legends constitute ultimate proof in themselves but, as has been noted by many scholars, it would seem peculiar that identical traditions should exist in lands separated by oceans or other vast distances without there being a grain of truth behind them.

The space prophet, Constantine Tsiolkovsky, was certain that a great deal of space traffic takes place between different solar systems, and he was confident that our Earth had not been neglected on the cosmic routes of interstellar travellers. In 1928 Tsiolkovsky wrote: 'It is difficult for us to imagine a being superior to earthman. This narrow-minded view prevents us from picturing an intervention of extraterrestrial entities in terrestrial affairs. Yet a great number of phenomena still remain unexplained because of this attitude. Many puzzling happenings are recorded in history and literature.'[8] The

[8] C. Tsiolkovsky, *Volya Vsellenoi* (The Will of the Universe), Moscow, 1928.

Russian scholar refers here to the appearance of angels and gods described in sacred books, classical writings and legends.

Picture a spaceship taking off from a planet of a highly civilised galactic race to travel towards a yellow star surrounded by a number of planets, among which might be a possible garden of rational life. Entering that solar system, the commander of the spacecraft directed: 'Destination—planet in third orbit from star. Object of mission—exploration and whatever scientific aid we can bring.' If this is mere fiction, then all the scriptures, history and folklore which have been so painstakingly accumulated and handed down by so many peoples for so many thousands of years, fall by the wayside.

15: *Search for the Tombs of Osiris*

The legend of Osiris and Isis is a typical culture-hero myth. In the beginning uncouth savages lived in prehistoric Egypt. On his descent from the starry heavens Osiris, with the help of Thoth, taught them the arts and sciences while his divine sister-wife initiated agricultural and cultural programmes.

Their civilising mission was hindered by the evil Seth-Typhon who enticed Osiris to lie in a coffin which was then nailed, sealed and thrown into the river Nile. When it reached the Mediterranean, the casket slowly drifted towards Byblos in Lebanon. In her sorrow Isis instituted a search for the body of Osiris, finally located the coffin and brought it back to Buto in Egypt. However, hateful Seth stole the casket with the body which he cut into fourteen or sixteen parts, scattering them over Egypt. Isis then collected them and put them together, lamenting: 'Return to thy house, Osiris.'

The entire legend might just be an allegory of the arrival of space visitors who brought with them a ready-made cosmic culture including artefacts of sophisticated science which the priesthood were told by the donors to preserve in secrecy. Accordingly, the fragments of Osiris's body would be the caches planted by the superior civilisers before they left. From the map designating the places where the scenes of the legend unfolded, might also be indicated the location of the hidden treasures.

An historical oddity from predynastic Egypt lends support to this idea. Osiris, son of Nut, the goddess of the starry firmament, was known as the god of the Underworld. This seeming contradiction of a god from the stars seated on a throne in the interior of the Earth, might be reconciled by the supposition that this is a figurative story depicting the descent of a celestial being who, during his sojourn on this planet, buried a treasure inside the Earth, which has still to be discovered. Over-imaginative as this interpretation might seem,

(Left) Osiris, son of the star goddess Nut, was the god of the Underworld. This may allegorise a cosmic civiliser planting underground caches of treasures. (Right) Map of places mentioned in the myth of Isis and Osiris might indicate their locations.

it could facilitate an archaeological discovery beside which even the tomb of Tutenkhamun would fade into insignificance.

As the drama of Isis and Osiris unfolds in Thebes in southern Egypt, this area must naturally be included in the chart of buried 'time capsules'. The parts of the dismembered body of Osiris were found by Isis in the delta towns of Athribis,

Bubastis, Busiris, Sais and Balamun. Then as Isis travelled southward up the Nile, she stopped at Heliopolis, Memphis, Faiyum Lake, Cusae, Siut, Abydos, Dendera and Elephantine (now Aswan). Since the whole body in the coffin was found the first time at Byblos, this site must also be included in the list as well as Buto where it was brought from Lebanon by Isis.

The priesthood was ordered by the visitors to construct eventually suitable markers, such as temples or pyramids, over the vaults. According to the legend, at each of the places on the map the priests were assured that what they were receiving was not just a fragment but the complete body of Osiris. This detail suggests the possibility of either a multiplication of information and artefacts or a network linking the sites. We can safely assume that the secret knowledge of the locations of the caches was handed down and confined to the priesthood over innumerable centuries, and that it was only at a far later period that the permanent markers were constructed. Also handed down through the priesthood were the instructions to guard the contents of the caches for ages to come, until such time as mankind should be ready to receive the cosmic treasures. Let us now scrutinise the map.

Elephantine or Aswan does not provide any clues to the caches. However, Thebes which was the city of Osiris and Isis, is more promising. The first indication of prehistoric science is located there. The chanting colossus of Memnon, referred to by classical writers and examined by Roman emperors, might be evidence of a scientific skill of high calibre. The statue itself is still intact, but the mechanism by means of which the colossus produced sounds at sunrise, was damaged during restoration work in ancient Roman times. Evidently, the device must have been too sensitive for the heavy incompetent hands of Roman engineers.

Near Thebes is situated the mortuary temple of the world's

first queen, Hatshepsut, at a place called Deir el-Bahri. It contains the burial chamber of Senmouth, her royal architect. The ceiling of this tomb has two astronomical charts and, in one of these, the cardinal points are enigmatically reversed, making it appear as though the Earth had shifted on its axis. Actually, the sky did look different over 12,000 years ago due to the precession of the equinoxes. This star map suggests that the learned priests kept records for untold periods of time. Obviously, the queen and her architect must have been aware of the true significance of this chart. There could have been no mistake on the part of Senmouth since three other ancient Egyptian documents—the Harris, the Ipuwer and the Hermitage Papyri, all corroborate this anomaly.

Next, at Dendera there exist ruins of a temple whose ceiling was engraved with a zodiacal chart. This Dendera Zodiac was brought to Paris in 1821 and is now displayed at the Louvre. In this chart the zodiacal signs are arranged in a spiral with the sign of Leo at the vernal equinox. Leo was in that position about 12,000 years ago due to the precession of equinoxes. Again, this is an indication of the existence of an extremely ancient calendrical system.

At Abydos archaeologists have uncovered the temple of Osiris which, according to Herodotus, had subterranean sanctuaries. This circumstance would be difficult to ascertain as the place is now flooded. But an impressive stone sarcophagus of Osiris has been salvaged at Abydos, and it is now on display at the Cairo Museum. Nothing definite can be said of Siut or Cusae until proper exploration is made.

Beyond Cusae lies Oxyrhynchus where, according to Plutarch, an oxyrhynchus fish in the Nile swallowed Osiris's penis. This may allegorise an accidental loss of a part of the Osirian treasures in the Nile river. No stone monuments or inscriptions of interest exist at Oxyrhynchus but two papyri, discovered by Grenfell and Hunt in 1897 and 1903, might have a bearing on our theory. Although they have been dated to the end of the second century, the scholar E. Hennecke writes that 'the text has been copied from an earlier original'.[1]

The *Oxyrhynchus Papyrus 654* reads: 'Let him that seeketh cease (seeking till he) find, and when he findeth (he shall marvel and) having marvelled, he shall reign.' The *Oxyrhynchus 655* has an intriguing phrase: 'The key of knowledge have ye hidden.'[2]

The Faiyum Lake district presents a site of unusual interest. Herodotus was impressed by the famous labyrinth near Lake Moeris. The buildings contained 1,500 rooms and an equal number of underground chambers which the Greek historian was not permitted to inspect. According to the priests numerous statues and scrolls were stored in these apartments.

Herodotus was informed by the initiated priesthood that the descent of the sky-gods had taken place in 17,500 B.C. and that this traffic continued until 11,850 B.C. after which 'no god ever assumed mortal form'. He was also told by the priests that Osiris (Dionysus) had appeared in Egypt about 15,550 B.C.

However, demigods apparently stayed with mankind much longer. According to Manetho, the Egyptian historian, whose list of pharaohs is still used in Egyptology, the first dynasty of demigods was inaugurated in 9013 B.C. The equally reliable *Turin Papyrus* dates the same event to 8813 B.C., that is 200 years later than Manetho.

Herodotus well realised the vastness of the period he was writing about as he stated that since the legendary era 'the sun had changed its usual position four times', obviously having in mind the precession of the equinoxes, already noted by us in the tomb of Senmouth. Speaking of the great antiquity of the

[1] E. Hennecke, *New Testament Apocrypha*, London, 1963.

[2] M. R. James, *Apocryphal New Testament*, Oxford, 1953.

golden age of the gods this Greek historian remarked: 'They (the priests of Egypt, A.T.) claim to be quite certain of these dates for they have always kept a careful written record of the passage of time.'

This bygone age of gods and demigods is, of course, incredibly remote but what is even more amazing is the fact that these pages from protohistory should have been recorded at all by the chroniclers of ancient times, if they had been merely fable.

When modern scholars write history books, they consult Herodotus, Manetho and the *Turin Papyrus* but they scrupulously avoid all mention of the celestial rulers of pre-dynastic Egypt and relegate this portion of their source data to myth. The logic of such an approach can be questioned.

The pyramids and the Sphinx at Memphis provide such a vast field of investigation that a separate enquiry is necessary. Nothing concrete can be offered as evidence of the presence of hiding-places in the delta towns of Athribis and Balamun as practically nothing remains of them. However, Sais, Bubastis, Busiris as well as Buto, where Isis landed on her return from Byblos, give indications of the existence of secret crypts or of an ancient scientific tradition.

Plato's *Timaeus* and *Critias* state that about 560 B.C. in the temple of Neith at Sais there were secret halls containing historical records which had been kept for more than 9,000 years. Proclus gives the name of the high priest with whom Plato spoke in Sais—Pateneit. It is probably from him that the Greek philosopher learned about the oldest archives of Egypt. Another interesting fact to notice is that the high priest of Egypt Psonchis, teacher of Pythagoras, also mentioned *sacred registers* which even speak of a collision of the Earth with a giant asteroid in a remote past.

Buto, the seat of the famous oracle of Leto, the mother of Apollo, was revered by Egyptians and Greeks alike. The shrine was built inside a colossal cubic stone 60 feet high and 60 feet wide. According to Herodotus, the oracle was 'the most veracious in Egypt'. If it was famous for correct predictions, it can be surmised that the priests had possessed the secret of breaking the time barrier, which would indicate mastery of an unknown science. The oracle gave a strange prophecy to prince Psammetichus, exiled by the other co-rulers of Egypt, that bronze men would make their appearance from the sea, bringing help. Psammetichus did not believe a word of it but not long afterwards Greek pirates were forced to land on the Egyptian coast. They wore bronze armour. The prince then recalled the words of the oracle and persuaded the raiders to enter his service. Thus he deposed his eleven enemies and became the pharaoh of Egypt—Psammetichus I. Buto, a winged cobra goddess, employed a symbolism which was as common in Egypt as it was in Australia or Central America—that of the *Plumed Serpent* which had brought to Earth the stellar benefactors.

Herodotus describes the festivals of Bubastis which were attended by as many as 700,000 people who came great distances in barges for the celebrations but, unfortunately, he does not disclose a great deal about his contacts with the priests. But when he mentions the great celebration dedicated to Isis held at Busiris, he writes that his reverence was so profound he did not feel it was appropriate to reveal what actually took place.

Nevertheless, the *Papyrus of Leiden* (first century) fills this gap. The manuscript was found in the tomb of Horsiesis, a priest and 'master of secrets', and gives an account of his pilgrimage to and initiation in the temple of Osiris at Busiris, or Per Ousir which means the 'sanctuary of Osiris'. This papyrus is evidently a ritual text as almost identically worded papyri have been found in other tombs of Egypt.

The *Leiden Papyrus* relates how Horsiesis walked through

dark corridors and then entered a crypt bathed in light, with seven doors:

The guides introduce thee
Into the most holy place.
In the divine crypt
Thou seest the sacred body
Lying on its funeral bed.

Although the text was used in funerary rites, it could at the same time be a ciphered message, for future generations, pointing to the time capsules buried in sacred places.

It is most significant that a relation exists between Dendera and Busiris, both shown on the map of the tombs of Osiris. A text on the wall of the temple of Hathor in Dendera refers to the crypt in Busiris:

As to the upper tomb (in Busiris),
The great god dwells there.
As to the crypt, it is made of stone.
Its height is 8·32 metres, its width 6·24 metres.
It has seven doors. There is one door in the West
Through which one enters.

This accurate description giving precise dimensions would appear to take it out of the realm of fable.

The Memphis area contains the three most prominent pyramids—those of Khufu, Khafre and Menkaure (Cheops, Chephren and Mycerinus). Traces of a cosmic science are present in the proportions of the Khufu pyramid. The base of its perimeter divided by twice the height gives π, thus displaying advanced mathematical knowledge. It is surmised that the ancient Egyptians had even attempted to solve the problem of squaring the circle.

In reality the Pyramid–Sphinx complex may be the site of a secret hall of records. This supposition is based on historical accounts which, however, do not furnish any clues as to the exact location of the hiding-place. The Roman historian Ammianus Marcellinus (fourth century) made these disclosures about the pyramids: 'Inscriptions which the ancients asserted were engraved on the walls of certain underground galleries constructed in the interior of certain of the pyramids, were intended to preserve ancient wisdom from being lost in the flood.'[3] This assertion made by a classical writer regarding subterranean vaults at Giza, is highly significant. Crantor (300 B.C.) stated that there were certain pillars or towers in Egypt which contained a written record of prehistory. Eusebius (fourth century) wrote that Agathodaemon (before 3000 B.C.) had deposited scrolls in the sacred libraries of the temples of Egypt.

There is hardly any doubt that the Giza Pyramid complex is paramount in our chart of Osirian treasure sites since, according to the sacred tradition of Egypt, the head of Osiris was buried at Memphis where the pyramids stand.

Some of the most fantastic accounts of the pyramids come from Arab sources. Abd el-Latif, an Arab author, referred to the thousands of heiroglyphs irretrievably lost when the pyramids' outer covering of polished limestone was removed to be used as free building material by the house builders of Cairo over 1,000 years ago. Balkhi, an astronomer-astrologer of the eighth century, also claimed that upon the exterior of the pyramids 'every charm and wonder of physic was inscribed'.

Masoudi (tenth century) in his story of the pyramids alluded to the Coptic tradition, which is interesting since the Copts are the only descendants of the ancient Egyptians. He contended that in a very distant past the pharaoh ordered the priests to deposit within the pyramid vaults 'written accounts of their wisdom and acquirements in the different arts and sciences'. This covered everything from history through geometry and

[3] *Ammiani Marcellini Rerum Gestarum Libri*, Leipzig, 1875.

medicine. He felt that the reason for hiding this time capsule was 'that they might remain as records for the benefit of those who could afterwards comprehend them'.

Masoudi also alleged that the Menkaure pyramid contained thirty secret repositories of treasures, including such extraordinary objects as 'iron which could not become rusty' and 'glass which could be bent', in other words—things undreamt of in the tenth century—noncorrosive iron, and plastics. According to Masoudi, the subterranean galleries of the three pyramids were guarded by mechanical statues of amazing capabilities, comparable to computorised robots programmed for discrimination, as they destroyed all 'except those who by their conduct were worthy of admission'. This is phenomenal information as it is possible that since the times of Masoudi 'worthy persons' have indeed seen the buried treasures.

The Arab scholar tells the story of twenty men exploring one of the pyramids in the hopes of finding something of value. One of the men fell into a deep pit, emitting horrible cries. His companions waited until evening when at last he struggled out of the sand shouting 'He who meddles with and covets what does not belong to him is dishonest.' On this utterance he fell dead and was carried away by his friends.

In the same century another writer, Muterdi, gave an account of an incident in a narrow passage at Giza, where a group of people were terrified to see one of their men crushed to death by a stone door which, of itself, suddenly closed the corridor in front of them.

Altelemsani, another Arab writer, states in his manuscript, preserved in the British Museum, that there is an underground passage between the Nile and the Great Pyramid. He relates the following episode:

> In the days of Ahmed Ben Touloun a party entered the Great Pyramid. They found in one of the chambers a goblet of

> glass of rare colour and texture. As they were leaving they missed one of the party and upon returning to seek him, he came out to them naked and laughing said: 'Do not follow or seek for me,' and then rushed back into the pyramid. His friends perceived that he was enchanted.

Upon learning of the happenings at the pyramid, Ahmed Ben Touloun expressed the desire to see the goblet of glass. During the examination it was filled with water and weighed. Then the empty cup was weighed again. The historian writes that it was 'found to be of the same weight when empty as when full of water'. If the chronicle is accurate, this lack of additional weight brings us again to indirect evidence of the existence of an extraordinary science.

Last but not least, a few words must be said about the Sphinx. This structure was incredibly old even to the ancient Egyptians. Its lion's body corresponds with Leo, the first sign on the Zodiac of Dendera. Upon the Sphinx's head a fragment of the cobra crown is still extant. It is notable that the pharaohs wore a double crown—a serpent for Northern Egypt and a falcon for the Southern Kingdom. In Mexico Quetzalcoatl was called the Feathered Snake, which implies an almost identical symbolism. The meaning might be interpreted in this manner—the bird-men built nests underground, as serpents, indicating the buried caches. The allegory is universal—the Nagas of India, known as the flying serpents, are believed to have built vast subterranean cities in the Himalayas.

A century ago A. P. Sinnett, a British journalist in India and one of the few Europeans ever to become a pupil of a Himalayan Mahatma, left a remarkable note about the Great Pyramid:

> I have gathered a hint to the effect that, although no doubt from the beginning used as and designed to be temples or chambers of initiation—the Great Pyramid, for one,

> certainly containing other chambers besides the three that have been discovered—one purpose of the Great Pyramid was the protection of some tangible objects of great importance having to do with the occult mysteries. These were buried in the rock, it is said, and the pyramid was reared over them, its form and magnitude being adopted to render it safe from the hazards of earthquake, and even from the consequence of submergence beneath the sea during the great secular undulations of the Earth's surface.[4]

Sinnett also mentioned that Pharaoh Khufu simply restored some portions of the pyramid, closing secret chambers and leaving his cartouches.

In the context of this theory of underground crypts, the X-raying of the Khafre pyramid assumes great importance. The Pyramid Project, a joint U.S.A.–United Arab Republic venture, intended to probe this pyramid with cosmic rays in order to locate hidden chambers or galleries, was originated in 1966 by Dr Luis W. Alvarez, a Nobel prize nuclear physicist. On the Arab side was Dr F. El-Bedewi of Ein Shams University of Cairo. The operations began in 1967 but were interrupted by the Middle-Eastern war. In passing through the mass of the pyramid cosmic rays are slightly handicapped by the stones, but if cavities are encountered, the rays travel faster and leave corresponding marks on the film. Numerous X-raying from two points can thus establish not only the presence of a secret chamber but can pinpoint its exact position. In 1969 Dr Amr Gohed was baffled by an interference in the probe which could not be accounted for, but Dr Luis Alvarez discards these rumours as unfounded. Since the cosmic-ray counter would have to be below the crypts to register their presence, in all probability they are hidden far too deeply below the pyramid to be detected by this type of probe.

[4] A. P. Sinnett, *Collected Fruits of Occult Teaching*, London, 1920.

Next comes ancient Byblos in Lebanon where the coffin containing the body of Osiris was cast up on the beach by the waves. As it came to rest on a sapling, the growing tree soon enfolded it completely. Out of this tree, with the chest inside, a pillar was constructed for the palace of the king of Byblos. Learning of this, the goddess Isis immediately travelled to Syria and upon finding it, returned to Egypt with the casket in which was the whole body of Osiris. This part of the myth may signify that a complete cache was deposited in Syria, now Lebanon. This conjecture is supported by Eusebius, Bishop of Caesarea (third century). As regards the source data used by the Egyptian historian Manetho, Eusebius says that 'he copied from the inscriptions which were engraved in the sacred dialect and hieroglyphic characters upon the columns set up in the Siriatic land by Thoth, the first Hermes'. Two facts are disclosed in this chronicle of Eusebius—firstly, that some detailed historical records were inscribed on columns in Syria, now Lebanon, and secondly, that these archives were compiled by Thoth, a heavenly messenger in a pre-deluge epoch.

Josephus Flavius, the Hebrew historian of the first century, describes how the ancients, afraid that 'their science should at any time be lost to men', engraved upon pillars 'their discoveries'. He insisted that these records were kept in 'the land of Syria'.

It is noteworthy that the only place besides Egypt where high pyramidal obelisks with gilded tops were constructed, is Byblos which is supposed to be 8,000 years old. The priests of the Nile Valley believed in the actual presence of sky-gods in the distant past of mankind. These beings could have arrived in spaceships in Egypt and Syria of which the obelisks were reminiscent in shape. This links Byblos with the myth of Osiris and places it among the prospective sites in which a prehistoric crypt may be located.

There are no apparent marker-monuments in Lebanon indicating an archaic cache, except Baalbek. Here too, might be the site where the 'chest of Osiris' is still preserved. There is not a single edifice in the Middle East from Turkey to Egypt of the dimensions of the gigantic Baalbek Terrace. The colossal platform, on which still stand the columns of old temples at an elevation of 920 metres, is all out of proportion to the size of the temples upon it. Some of the slabs weigh close to 1,000 tons and no crane has yet been built—even today, capable of lifting these masses from the quarry. The colonnades and walls of the Roman temples of Jupiter, Venus and Bacchus are quite impressive, yet they truly fade into insignificance when compared with the base upon which they stand. Surveying Baalbek from a distance, it would seem that the platform had been built in a far earlier period and for other purposes than temples. Mark Twain visited Baalbek and wrote these words about the enormous foundation stones: 'They remain for thousands of years an eloquent rebuke to such as are prone to think slightingly of the men who have lived before them.' The temples on the colossal terrace and the poor villages below are such a contrast that Mark Twain said: 'They look strange enough in such plebeian company.'

According to the writers of antiquity, the mysteries of the temple of Hadad, or Jupiter, included such marvels as luminous stones and thunderbolts. This implies the manipulation of electricity or some other powerful energy, another evidence of an archaic scientific legacy. The great age of the Baalbek tradition can be gleaned from ancient writings which record the verbal tradition of Baalbek priests in Roman times to the effect that the edifice was erected soon after the legendary Deluge. Thus it is not at all improbable that the megalithic structures, whether at Baalbek or Giza, are merely permanent markers indicating the location of the treasures of celestial visitors. Until archaeologists explore all these sites and uncover just one of these crypts, this basic theory must remain in the realm of fantasy.

Nevertheless it would not be superfluous to cite a legend which the author happened to learn in Lebanon. It tells of the killing of Abel by Cain in the vicinity of Baalbek's megalithic foundations under which Abel's heart finally stopped in the underground labyrinth under the massive stones of Baalbek. This lore about a labyrinth under the terrace of Baalbek stimulates speculation as to possible caches buried deeply under the giant blocks.

There exists another source of information as regards this hypothesis—the testimony of the Ancient Mysteries and of secret fraternities that perpetuate the arcane tradition. Lucius Apuleius (second century), a priest of Osiris, writes about his initiation into the Mysteries of Isis which could well have taken place in the Great Pyramid, or in a secret temple underneath it: 'At midnight I saw the sun shining as if it were noon; I entered the presence of the gods of the underworld and the gods of the upperworld; I stood by and worshipped them.' It is quite certain that he could not have seen the sun at midnight—at least not in Mediterranean regions. But if there had been a powerful generator and brilliant artificial lighting in the subterranean shrine, he could have spent some time there in the company of projected images of gods from another solar system.

The Freemasons preserve a tradition of 'winding stairs and hollow pillars' which are 'receptacles of the archives of the Craft'. The Royal Arch Masons still perpetuate the 'secret vault of King Solomon' and 'the way of Hidden Treasures' as well as the cavity under the Keystone—which might allude to the Great Pyramid.

A description of an underground temple containing a cache is, perhaps, included in the *Fama Fraternitatis*, the Rosicrucian manifesto of 1614. The adepts who issued this document offered their science to worthy persons, 'to prince and peasant' alike, promising that 'the rule of false theology shall be overthrown' and that 'contradictions of science and theology shall be reconciled'.

To equalise prince with peasant as much as to condemn false theology was outrageous at the time. In that century of feudalism dominated by the Church, it is no wonder that those adepts admitted in the manifesto that they had to 'remain concealed'. Ever since the publication of this undoubtedly ciphered document in 1614, that is immediately after the execution of Giordano Bruno in 1600 and during the persecution of Galileo, modern science began its assault on the prejudices of the period. The coincidence between the date of *Fama Fraternitatis* addressed to the 'learned of all Europe', appealing for a world reformation, and the actual beginnings of modern science and democracy, has never been explained.

The *Fama Fraternitatis* describes the opening of the tomb of Master Christian Rosenkreuz, which might well have some parallel with the subterranean crypt of the Giza pyramids. The sepulchre had seven walls with doors leading to storehouses of books and scrolls. A circular altar stood in the centre, covered with hieroglyphics. The vault was brilliantly illuminated by an artificial 'sun' in the ceiling. This may be a carefully worded account of a visit by an adept to the secret depository at Giza. De Montfaucon de Villars gives an interesting detail in this story of the opening of the vault of Christian Rosenkreuz.[5] There was a 'statue in armour' (or robot) which destroyed the source of light when the chamber was opened. This is strangely similar to the accounts of the Arab historians who claimed that automatons guarded the galleries of the pyramids.

Turning back to Baalbek we find an esoteric science among the Druses of Lebanon. They believe that at the time of the

[5] Montfaucon de Villars, *The Diverting History of the Count of de Gabalis*, London, 1714.

descent of Hakim, a sky-being, the world was 3,430 million years old. Since the age of the Earth is estimated at about 4,500 million years, it follows that the event must have taken place about 1,000 million years ago. From where did the Druses procure this fabulously ancient calendar—and could it be connected with the cache of the star-born?

It would be appropriate to indicate here that according to a Druse tradition the name of every Druse is inscribed in the unknown vaults and galleries of the pyramids of Giza which suggests a possibility that the initiates know their secret entrances. The Druses are certain that the pyramids likewise contain papyri and precious objects. During my stay in Lebanon Prince Kamal Joumblat, a Druse initiate and chief, mentioned the sacred records of the brotherhood predicting the discovery of the pyramid treasures before the year 2000.

This chapter represents an irrational attempt to uncover a rational pattern in the legends of antiquity, disclosing the location of buried treasures of those who came to Earth from the stars thousands of years ago, leaving gifts for a mature mankind of the future. The contents of these archives and crypts may reveal their builders' knowledge of forces which are utterly new to us. They may also contain records going back for tens of thousands of years to an era when gods or angels trod the Earth. These conclusions can only be verified by a systematic search by qualified scientists for the caches of the visitors from space. If this discovery is ever made, it will create a revolution in science, the like of which has never even been imagined.

16: *Labyrinths and Serpents*

The Minoan civilisation of Crete, one of the most ancient in the world, began around 2500 B.C. or about the time of the reign of Pharaoh Khufu. Crete became a great maritime power by trade rather than by war, and with its commercial links all over the Mediterranean, the Minoan empire achieved great prosperity. It did not have any walled cities which proves that the kingdom was united and the population homogeneous. Isolation by the sea helped to preserve this peaceful country till 1400 B.C. when suddenly it was wiped out by some unknown disaster.

The Cretan religion was based on the cult of the Universal Mother and Snake Goddess, combined with the worship of the bull, the symbol of strength and creative energy.

Knossos had a five-storied palace covering an area of 16,000 square metres. The drainage and ventilation systems were almost modern. Everything spoke of high development. Daedalus, the architect of the palace of Minos, also dabbled in aviation. When he and his son Icarus were forced to leave Crete, they used gliders. The father flew safely to Sicily but Icarus fell into the sea.

According to legend, King Minos of Crete ordered his architect Daedalus to construct the labyrinth, a maze of passages so ingeniously devised that even the builder himself could not find his way without a plan. In the centre lived the Minotaur, half bull, half human, to whom the Greeks sent seven youths and seven maidens as a tribute every nine years. The Minotaur was slain by Theseus who was able to find his way out of the labyrinth thanks to a ball of thread given to him by Ariadne. This myth has been interpreted as an historical record of the construction of the palace of Minos in Knossos which contains innumerable galleries and rooms. On the other hand, this myth may have an entirely different interpretation, similar to a cryptogram, which conceals the existence of a secret repository of underground chambers and passages. The labyrinth legend could be another clue as to the location of the

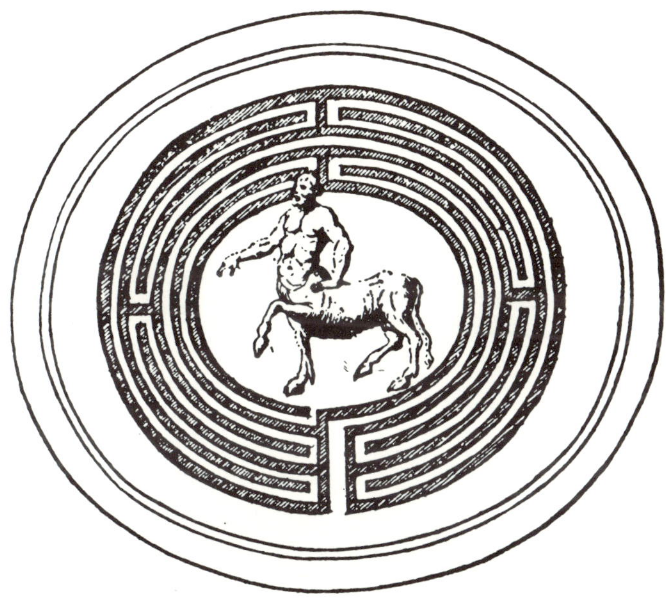

Legend of Minotaur and Labyrinth may be coded chart showing location of secret 'time capsules' buried at dawn of history.

buried 'time capsules'. It may well be that the pattern of the maze indicates the spot where such a treasure is hidden. Accordingly, it would be interesting to examine all Minoan mazes in view of finding a thread. The Minotaur, half bull, might be an intimation of artefacts stored when, in the precession of equinoxes, the sun entered Taurus, the Bull, around 4000 B.C., that is at the dawn of Egypt's history.

At Phaistos, Crete, there is a ceramic disc with a circular

maze of thus far undeciphered hieroglyphs. The disc, estimated to be about 3,700 years old, may be only an ordinary message. Yet the spiral arrangement of the characters or words, in the form of a labyrinth makes it intriguing. This possible flight of fancy is prompted by the fact that a scientific hypothesis must get outside the framework of the data which led to its creation in order to justify itself, by collecting the ultimate proof.

Labyrinth on silver coins from Crete (2nd century). Cryptogram indicating location of hidden caches left by wise men of antiquity?

Over 600 years ago Chow Ta-Kwan, a Chinese government official, presented to his emperor a written account of a fabulous city in Indo-China. When the document was translated and published in Europe in 1858, historians took no notice of it. About the same time a French naturalist Henri Mouhot, collecting specimens in Cambodia, accidentally heard of a lost city from native guides. He travelled through swampy jungle to discover Angkor Thom, an immense metropolis lost in tangled tropical undergrowth that still had traces of former roads, bridges and temples. This proved to be the city described in Chow Ta-Kwan's scroll. The kingdom of the Khmers appeared around the beginning of our era and lasted for fourteen centuries. The first inhabitants of this empire were serpent worshippers. Then followed an era of Brahminism, superseded by Buddhism, and eventually all amalgamated.

Chinese chroniclers state that a treasure is buried at Angkor Thom in a temple crypt. They also say that it is guarded by a huge image of Buddha. The Cambodians believe that a white python eternally watches the door to the crypt and will destroy any treasure-hunter or profaner who approaches it. Fantasy? Not quite. During the fifties before the escalation of the war in Vietnam, treasure hunters roamed around Angkor searching for the treasure. Two Englishmen are reputed to have reached the vault. One was found unconscious near Angkor Vat with a precious stone in his hand. When he recovered he could not give a coherent story of how he and his companion entered the temple and what they saw there. He was suffering from total amnesia but his friend was even less fortunate as he was found dead, crushed by a stone door.[1]

Similar tales of serpents guarding treasures circulate in South America, as for instance, the legend of the snake-god Urcaguay, the protector of a buried treasure. According to folklore, the ancient city of Cuzco was founded by four brothers, one of whom lived inside the rocky mountains. There is also a strange local belief that Cuzco and Tiahuanaco were once connected by a tunnel over 300 kilometres long. It was likewise said that at a certain time of day when the rays of the sun fell on one of the walls of the Temple of the Sun, symbols

[1] *Sydney Morning Herald*, 11 February, 1954.

and lines appeared on the wall. These were supposed to have formed a map of the labyrinth.

The latest confirmation of the South American legend of subterranean labyrinths comes from Erich von Däniken who describes in his *Gold of the Gods* the discovery of a vast network of underground passages by Juan Moricz in 1965. The walls of these galleries are polished and the corridors form perfect right angles, suggesting an advanced technology applied to their construction. According to local Indians who guard the treasure caves, this system of tunnels extends for hundreds of kilometres.

Metal and stone tablets, found inside the caverns, are covered with signs and unknown hieroglyphs. Some gold plaques, engraved with the emblems of the sun, moon, stars, pyramids and serpents flying in the heavens, are on display at Cuenca, Ecuador. Ethnologist Moricz claims that the metal tablets 'form a veritable metal library which might contain a synopsis of the history of humanity, as well as an account of the origin of mankind on Earth and information about a vanished civilisation'.

Legend has it that a vast subterranean network exists in the Andes. This would infer engineering skill of the highest calibre which the early inhabitants of South America did not lack. The megalithic stonework of Tiahuanaco has to be seen to be believed. Stones are fitted together with insets without cement as if they were ivory and not 20-ton stone blocks. These ruins were ancient even at the time of the Conquest, nor could the Indians of the period provide an answer as to the identity of the Titan-builders. Another puzzle at Tiahuanaco is the absence of burial sites in the complex.[2] Does the reign of the Tiahuanacan Empire, established by the Sons of the Sun, go back to an unknown chapter of prehistory?

[2] A. Kondratov, *Tainy Trekh Okeanov* (Secrets of the Three Oceans), Leningrad, 1971.

Tiahuanaco on Lake Titicaca is situated at a higher altitude than the top of eternal snow-capped Mount Fuji in Japan. An astonishing discovery was made on the shores of this lake—that of a quay with facilities for boats large enough to sail the ocean. All around were found fossils of seashells and seaweed. It is obvious that the age of Tiahuanaco has been greatly underestimated for the evidence would suggest that it had existed before some cataclysm elevated that portion of South America raising a seashore to an alpine level.

Some 40 years ago Professor Arthur Posnansky and Dr Edmond Kiess made an attempt to decipher the markings on the Gate of the Sun in Tiahuanaco. Their findings were hard to swallow—on that calendar the day is of 30 hours and the year of 290 days! In addition, there are recorded solar eclipses at frequent periods—19 eclipses every 24 days of the Tiahuanacan month. What is this—a calendar from some long-lost, forgotten era or one from another planet?

As if this enigma were not enough, when the Argentine Submarine Federation explored the bottom of Lake Titicaca it detected architectural ruins extending for a kilometre, with paved streets laid out in strict geometrical order. Perhaps this is the first city founded by the first Son and Daughter of the Sun.

The Jesuit Agnelio Oliva (1572–1642) recorded the words of an old Inca quipu reader to the effect that the real Tiahuanaco was a subterranean city exceeding the one above ground in vastness. It was believed that the entrance to the underground apartments could be gained through four tunnels. Last century one passage was evidently found as treasure hunters managed to get in, to look for gold, but only one came out. He brought out with him two gold bars but left behind his sanity. After this incident the Peruvian government decided to wall up the cave entrance.[3] How like the experience of the two Englishmen, treasure-hunting at Angkor.

[3] H. P. Blavatsky, *Isis Unveiled*, New York, 1886.

Tiahuanaco is known for its cult of the Weeping God. This seems to be a South American version of the Osirian myth of Egypt. Manco Copac and Mama Ocllo, the culture-bearing couple of South America, who descended from the sun, are counterparts of Osiris and Isis.

Americanologists believe that the Nasca culture on the Pacific coast is an offshoot of the Tiahuanaco civilisation. At Nasca there are extraordinary tracings on the desert which extend for over 100 kilometres. Due to oxidation the desert is of a darker colour than the soil underneath. This vast network of lines, geometric patterns and figures was created by removing the upper layer exposing the lighter undersoil. The project must have taken decades to complete. As the designs can be seen only from high in the air, their purpose defies most theories. Is this a calendrical system, signals for prehistoric aircraft or coded instructions on locating the caches buried by superior beings who, like Quetzalcoatl, flew in their serpent sky-ships?

And now a return to Asia. Even in this jet-age every Hindu is familiar with and usually believes in the legend of the Nagas, the 'serpents' which live in extensive underground palaces in the rocky Himalayas. It is believed that these creatures are able to fly in space and that they possess amazing magical powers and intelligence. They are not too fond of man if he is a curiosity seeker, explorer or mountaineer. According to the sacred tradition of the Hindus, the deep caverns of the Nagas contain fabulous treasures, illuminated by flashing precious stones. The subterranean abodes are known to be in certain parts of both the Himalayas and Tibet, particularly around the Lake of the Great Nagas—Lake Manasarowar.

What this legend of India portrays may be a cosmic base constructed by beings from a world, very different from ours, who created artificial shelters suitable for their organisms. Why are the serpents that are alluded to in holy books considered to be wise? Actually, it has been proven by experimentation that snakes are not clever at all. The epithet of the wise serpent might simply be the allegorical interpretation of cosmic beings who, after arriving in 'serpents' or 'dragons' from celestial spaces, chose to live underground in deep catacombs in the manner of snakes.

What has been hinted at so far, may not be readily accepted by those who, sitting in their comfortable soft armchairs pass judgment on a subject totally unfamiliar to them. But the explorers who have climbed the rocky, icy Himalayas facing bucking winds that could blow them off a ledge, who have marched across the arctic plateau of Tibet or trekked through the sandy wastes of the Gobi Desert through cold sandstorms that turn the sun red and the sky brown, might have a different view from that of the sceptic in the armchair. There are many—Cacella, Csoma de Köros, Przhevalsky, Kosloff, Grunwedel, Francke, Blavatsky, David-Neel, Ossendowski and Roerich. Many of these personalities have made definite statements confirming the existence of hidden treasures of the stellar civilisers.

Nicholas Roerich and Alexandra David-Neel, the noted orientalists, both wrote of Gessar Khan's prophecies: 'I have many treasures but only upon the appointed day may I bestow them upon my people. When the legions of Northern Shambhala shall bring the spear of salvation, then shall I uncover the depths of the mountains.'

In Sikkim Roerich was told of tunnels and giant caves that were used for storing ancient reliques. The mountain Kinchinjunga was so named because in Tibetan that means *Five Treasures of the Great Snow*. According to the lamas: 'the giant gate of this storehouse will one day be opened'.[4]

In Karakoram Pass (altitude 6,000 metres) at the western end of the Himalayas, Nicholas Roerich was told by a guide:

[4] N. Roerich, *Heart of Asia*, New York, 1929.

'even we lowly people know that there are deep extensive underground vaults in which are gathered treasures from the beginning of the world'.[5] During his expedition Nicholas Roerich and his son Dr George Roerich, professor of Oriental languages, obtained information from the lamaseries about hidden passages under the Dalai Lama's palace, the Potala, and about a grotto under the main temple. He recorded the legend of the black stone of Shambhala which allegedly came from another planet.

Helena Blavatsky spent at least three years in Tibet, Bhutan and Sikkim. Her encyclopaedic books contain a wealth of data on Asiatic lore:

> Along the ridge of Altyn Tagh whose soil no European foot has ever trodden so far, there exists a certain hamlet, lost in a deep gorge. It is a small cluster of houses, a hamlet rather than a monastery, with a poor-looking temple in it, with one old lama, a hermit, living near by to watch it. Pilgrims say that the subterranean galleries and halls under it contain a collection of books, the number of which according to the accounts given, is too large to find even in the British Museum.[6]

In another passage Blavatsky states: 'Built deep in the bowels of the earth, the subterranean stores are secure, and as their entrances are concealed, there is little fear that anyone would discover them, even should several armies invade the sandy wastes.'

All these records point to the startling possibility that a stellar race not only planted priceless scientific artefacts in widely distributed underground storehouses, but also appointed trusted priests, monks and scholars to guard them from generation to generation. This heritage could have been handed

[5] N. Roerich, *Himalaya—Abode of Light*, London–Bombay, 1947.

[6] H. P. Blavatsky, *Secret Doctrine*, Vol. 1, Adyar, Madras, 1938.

down in an epoch the memory of which only mythology preserves. In speaking of the similarities of the cataclysm myth, Larousse Encyclopedia of Mythology[7] raises a question: 'Are all these legends a confused account of great events on a planetary scale which were beheld in terror simultaneously by the men scattered everywhere over the world?' This planetary

According to lamaist beliefs, artefacts from other planets are allegedly preserved in such cave repositories. (*Power of the Caves* by Nicholas Roerich, Gorky Art Gallery, U.S.S.R.)

disaster may rightly be identified with the Great Flood and the sinking of Atlantis. And as many myths assert, the descent of the gods could have occurred about that time. So far as the scientific and historical treasures are concerned, these could have been concealed by super-beings and their terrestrial agents in several places, especially when disasters threatened whole civilisations.

It would appear from ancient and modern sources that the

[7] Larousse Encyclopedia of Mythology, London, 1960 (p. 455).

collections in these secret repositories in Asia have been regularly supplemented with books and works of art by their guardians. In fact it is rumoured that some of the paintings of European masters have found their way into these hiding-places. During his stay in Orvieto, Italy, Nicholas Roerich learned of an historical chronicle about the buried masterpieces of Duccio di Buoninsegna (1255–1319) or of one of his school. Because of continuous criticism of his work, the Italian painter decided to hide some of his works from the public eye. 'I shall collect the rejected paintings and bury them in the deep basement of a monastery', he was reputed to have said.[8] No one knows exactly where the art treasures were buried, but Roerich felt that the works of Duccio and of other master painters were secreted in a far-off land. If this is true, the masterpieces in the catacombs of the Himalayas and of Central Asia should be safe.

Blavatsky was of the opinion that many scrolls from the Alexandrian Library escaped the fire started by the Roman army because they had been dispatched to a secret museum in Tibet. She was quite certain on that score and substantiated her statement by saying that the guardians of these concealed libraries 'could if they chose, lay claim to strange ancestry, and exhibit verifiable documents that would explain many a mysterious page in both sacred and profane history'.[9] It also seems reasonable to suppose that the Essenes of the Qumran community could have been a minor centre of this kind. Who knows, one day we might be studying documents even more significant and exciting than the Dead Sea Scrolls.

Darjeeling in the Himalayas is a place not totally devoid of the atmosphere we sense in Kipling's India. There I met an educated man from Sikkim who spoke of the erudition of the high lamas of the Tashilumpo Lamasery in Shigatse. In its towering temples tuition in ancient science has always been reserved

[8] N. Roerich, *Nerushimoe* (The Indestructible), Riga, 1936.

[9] H. P. Blavatsky, *Isis Unveiled*, New York, 1886.

for the few promising pupils. He admitted that only a limited number of lamas knew anything about the mystery of snowy Kinchinjunga which at that time floated above us like a mirage.

Since this well-travelled man had visited many Buddhist monasteries, I asked him if he had known anyone who had visited the mountain crypts, and the following dialogue ensued:

He: Over the centuries lamas, gurus and even some Europeans have been inside them. But although they had seen many things, they said few words about them.

I: How did they get in?

He: No one can enter without a guide or a pass which is a coded map. But the *Om Mani Padme Hum* carved on rock in a special style is sometimes a sign that the door is near.

I: I suppose it would be very difficult to open a stone door that is used only every twenty years or so.

He: Strange as it may seem, that is not the case. An old lama told me that the door just slides inside with the greatest of ease as if on oiled rollers. However, the passage beyond the entrance is usually barred by a sheet of cold bluish fire.

I: That is interesting—is it something like the artificial lightning of a million volts produced in our laboratories?

He: All they say is that it is frightening yet fascinating to look at. They must walk through it. If they are properly trained they succeed, if not, they die. I would not try. But you must know about the firewalking and how people are not touched by heat that melts iron bars?

I: Yes, I do. But it is all the more surprising to hear that because a friend of mine who visited Giza many years ago, told me about a similar barrier of flames in a subterranean gallery under the Sphinx where he had been taken by an Arab member of a secret fraternity.

He: Wonders are many yet few are they who can understand them.

A few weeks later I decided to go to the Kulu Valley in the Western part of the Himalayas to visit Naggar, where Nicholas Roerich had lived. Since I had known him personally, the trip had a sentimental overtone. A narrow curving road, a precipice on one side with rocks and avalanches on the other, were not conductive to an enjoyable journey to this remote region near Ladakh and Tibet. The village of Naggar derives its name from *Naga*, the serpent. High up in the mountains lies Roerich's estate. Having been an artist of note, his two-storied house contains a museum of his paintings.

As I began my ascent on the mountain path, I saw a tall grey-haired sadhu (hermit), sitting by a mountain torrent. In his hand he held a cobra-shaped staff, which together with the markings on his forehead, signified that he was a devotee of Shiva. During the earlier, more peaceful times of the British Raj, these pilgrims would travel to the Lake of the Great Nagas, Lake Manasorowar, or to Mount Kailas, the abode of Shiva, in Tibetan territory. I climbed the mountain and reached the terrace on which the Roerich's house is built. I spent at least an hour studying the master's paintings. On the way back I admired the narrow valley and the looming snow-capped mountain ridges on both sides.

The sadhu was still there. I thought, 'A place called Naggar, a devotee of the Nagas with the cobra staff, if he does not know something about the Nagas, then who does?' Knowing the ways of the East, I saluted the holy man with the folded hands in the fashion that is customary in India, and waited for the older man to speak first.

'You like Roerich's paintings?' he said in fluent English.

'Very much, indeed. Tell me, did you know the master in life?'

'Yes, for many years. A great Rishi* and a friend of Nehru.'

* An inspired sage.

'Venerable sadhu, I believe in the Nagas. Have you seen them?' I asked diplomatically.

'I am a poor sadhu, I know nothing, sahib. But about twenty years ago my yogi teacher went into the mountain kingdom of the Nagas. Bright light everywhere, big halls like Taj Mahal. Wonderful. The Nagas have many, many things and machines. They are clever, like Cambridge men, may be more clever, sahib,' the sadhu said with an apologetic smile. I could not help laughing.

'Your yogi must have been a Rishi. Don't the Nagas destroy men by their sting?' I asked.

'Yes, though the Nagas are gods and wish nothing but good to man, they do not like men who have no business near their palaces,' he replied.

'If you ever see the Nagas, give them my greetings,' I said before parting. The sadhu tilted his head sideways three times, which in India replaces our nod. 'China lets in no more pilgrims, I can go only through the long holes but I am too old now,' he concluded.

Walking towards the village I looked back at the old sadhu with the cobra staff. My two encounters remain fresh in my memory—the one under the silvery glaciers of Kinchinjunga, the other by the torrent of Naggar, on the bank of which sat that ascetic of Aryavarta, the holy land of the Mahabharata.

A description of the contents of a Tibetan subterranean museum is given by C. W. Leadbeater in his book *The Masters and the Path*. Doubts as to its authenticity may be partly justified not because the evidence itself is false but due to its naïve presentation. Leadbeater claims that the museum contains statues of the different racial types that go back to the beginning of time, the profiles of continents with their changes, diagrams of ethnical and religious fusions, and much more besides. There are, he says, 'strange scripts from other worlds than ours'. This is an exciting declaration as it implies that communication with other planets took place in the far-distant past of which our Western science is unaware.

It is obviously impossible to verify these claims since the general public is barred from entrance to the prehistoric museum. However, in the course of my travels I have been shown sketches drawn in the Chinese style depicting a crypt of this kind. These brush drawings included statues of human giants standing in a cave lit by the torches of lama-students only half as tall, but these were less astounding than the painting of a cigar-shaped aircraft (or possibly spaceship) horizontally moored on a conic tower. I am still puzzled as to how and why it kept its balance in such a position. But the cone was also apparently used for climbing up to and into the vessel. It was impossible to get information about its propulsion but the size was estimated to be approximately that of today's jet airliner with no less than fifty seats. In view of the episode of these brush drawings I hesitate to write off Leadbeater's story.

In his painting *Power of the Caves*, displayed in the Gorky Art Museum of the U.S.S.R., Nicholas Roerich portrays a cave repository of this type. This painting depicts entrances to spacious caverns guarded by lamas. There is little doubt that Roerich actually did visit this ancient museum as was confirmed in a conversation with a member of Roerich's Trans-Asiatic Expedition while I lived in China some 35 years ago. The title of the picture is appropriate, as knowledge is power, and great knowledge is hidden in those caves.

In recapitulating this chapter it appears that a great similarity pervades the folklore of many countries no matter how far apart they are. Traditions of vaults, labyrinths, tunnels and buried treasures of remote antiquity are found in Crete, Egypt, Tibet, Angkor, India, Mexico, Ecuador, Bolivia and Peru. The legends usually connected with the cult of the Serpent come especially in Egypt, Crete, Angkor, Tibet, India and Mexico. The fables of star-gods who came to Earth to civilise mankind, leaving hidden treasures for future civilisations, are prevalent in the myths of Egypt, India, Tibet, China, Mexico, Bolivia and Peru. That there are mechanical or human guardians preserving the safety of these storehouses of artefacts from another world as well as from bygone ages, is mentioned in the scriptures and lore of India, Tibet and Egypt.

No doubt there has been a considerable amount of theorising here, but how can truth be ascertained unless someone has the courage and impertinence to explore unknown territories? Progress in fresh fields of science has always been accomplished by the bold pioneering approach which ignores ridicule, failure and disappointment. This research may possibly furnish an Ariadne's thread by means of which modern science might be capable of finding its way in this historical labyrinth, and thus locate the cultural and scientific treasures of those wise super-beings who ruled mankind in a golden age so long ago.

17: *Traces of the Star-born*

The supposition that astronauts from another world in space could have left artefacts on Earth, has been advanced by noted scientists. Professor F. D. Drake of Cornell University agrees that these visits of extraterrestrials could have taken place in the past: '. . . they would come so infrequently, the number of artefacts would be very small, and they would be buried perhaps now in archaeological ruins.'[1] He has also suggested using radiation detectors in order to locate such caches, planted by interstellar visitors, which might be tagged with radioactive isotopes of an artificial pattern. The monolith in Kubrick–Clarke's *2001—A Space Odyssey* illustrates dramatically what might be a true gift from the stars.

In fact such a cosmic artefact may have been found by Lunokhod 2 in the Sea of Serenity in the foothills of the Taurus Mountains on 14 February, 1973. A one-metre-long stone monolith of uncommon strength and smoothness was examined there by this Soviet probe. The stone plate had an unusually smooth surface unlike the large lunar stones lying near by and, surprisingly, it was dated to a much more recent period than the surrounding rocks of the moon.[2]

There may still exist traces of galactic astronauts' sojourns on Earth which elude identification. Certain objects that are wrongly identified in museums may belong in this category. Only one genuine souvenir of the space visitors would be required to validate this theory. Until this programme is

[1] J. Agel, *The Making of Kubrick's 2001*, New York, 1970.

[2] *Soviet Aerospace* (1341 G. St., N.W., Washington, D.C., 20005), 19 February, 1973, p. 56.

carried out in depth by an international scientific body, the search for these articles may remain largely futile.

Meanwhile, the reader is invited to go on an imaginary expedition in pursuit of the vestiges of stellar explorers on this planet. Persuasive evidence is available that the space Vikings have left footprints that have been well preserved over the millenia. Outlines of hands, and even portraits were skilfully rendered by the primitive tribes who had witnessed their landings.

In 1959 a Soviet–Chinese paleontological expedition, under the leadership of Dr Chow Ming Chen, brought from the Gobi Desert an unusual artefact that appeared to be a fossilised imprint of a ribbed sole in sandstone.[3] It was not difficult for the members of the expedition to date the geological layer in which it was found but identification of the find completely escaped them. The sedimentary rock was formed by a sea in Central Asia that had existed there more than 15 million years ago, before the elevation of the Himalayas and the Tibetan plateau.

The team had enough specialists to be able to recognise that this was not the fossilised paw of some prehistoric beast but the print of an apparently modern shoe, and to add further complication, this specific type of ripple-soled footwear worn millions of years ago suspiciously resembles the space boots of the Apollo astronauts who left similar tracks on lunar soil. No palaeontologist has yet ascribed any satisfactory explanation to this sandstone cast. Oddly enough, according to ancient Brahmin belief, the 'Kumaras' descended on an island in the Gobi Sea millions of years ago. What correlation is there between the footprint in the Gobi Desert and this legend?

Human hand-prints are often found among the rock paintings of Australia. To the aborigines they have a ritualistic significance. These 'hand shadows' serve as signatures of the human souls

[3] *Smena*, No. 8, 1961 (U.S.S.R.).

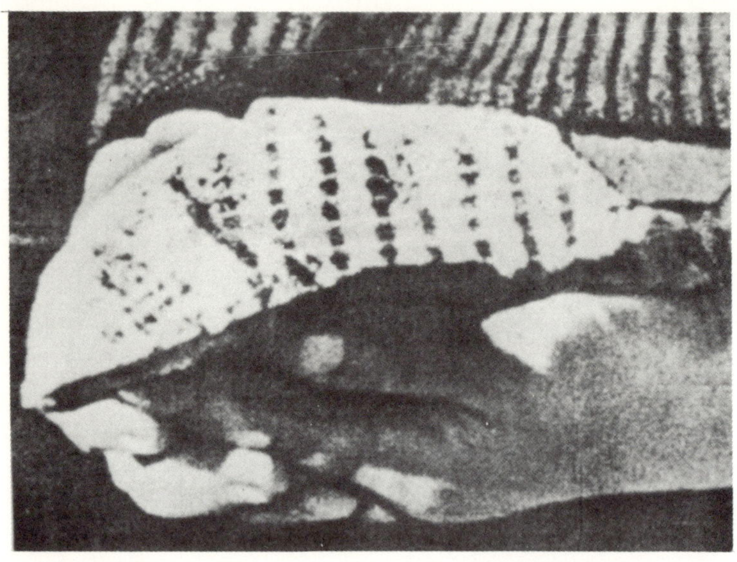

(Above) Fossilised ribbed-sole on sandstone, millions of years old, found in Mongolia. (Below) American astronaut's footprint on lunar soil.

that will outlive their possessors and will thus become part of the ancestor cult. Certain cave paintings on granite found at Garnet Glen, Mann Ranges in Central Australia, depict the

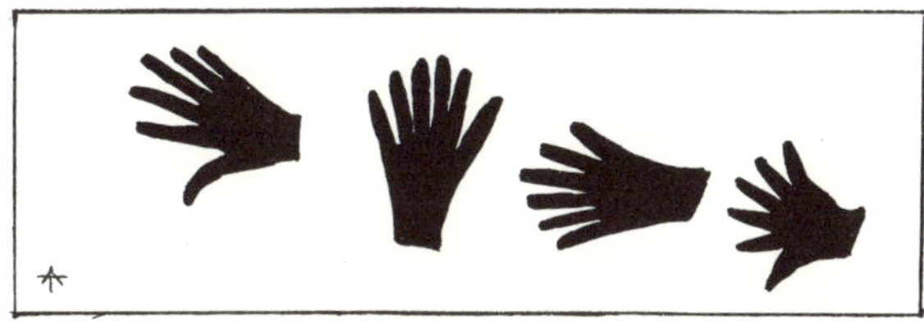

(Above) *Rainbow Serpent* rock paintings in the Kimberleys, Australia. Descending spacecraft, spiral flight orbit, astronauts in spacesuits? (Below) Prints of six-fingered hands, drawn by the aborigines, may belong to beings of other worlds.

silhouette of a hand with six fingers. It is included in the context of symbols connected with the *Dreamtime* which, according to Australian mythology, was the legendary epoch of the first men. There are symbolic aboriginal patterns at Eucolo Creek,

Pimba in South Australia which also clearly display a six-fingered hand.

The giant Wandjina rock paintings, that represent some of the most ancient specimens of the primitive art of that continent, depict mythological sky-beings having hands and feet with as many as seven fingers and toes. No animals or men on earth

A typical six-fingered figure, carved on stone slabs, found on Marcahuasi Plateau and in Casma Valley, Peru.

are known to have had that many appendages. These drawings are so old that even the aborigines themselves disclaim their ownership. 'It happened in the Dreamtime when gods fought giant monsters,' they say, thus alluding to a prehistoric era. Did Wandjina culture-bearers from the sky have more fingers and toes than earthman?

Australia is not the only place in the world where such odd hand-prints are found. At an altitude of 4,000 metres on the

Gold sun god with four fingers and four toes on hands and feet. (Cuenca, Ecuador.)

Marcahuasi Plateau, 80 kilometres from Lima, Peru, Dr Daniel Ruzo made the interesting discovery of a 'museum' of rock sculptures in 1952. Among them, the sculpture of a human figure in a helmet has six-fingered hands, and Dr Ruzo named it *La Mano*. This strange carving is not unique in South America.

Carved on gigantic slabs of stone at Cerro Sechin, Casma

Valley in northern Peru are many similar figures with six-fingered hands which astonish all who examine them.[4] The vast network of lines in the Nasca Desert of Peru, called the eighth wonder of the world, also contains a huge figure with six-fingered hands.[5] In view of the fact that these patterns in the desert can be seen only from a high altitude, this gives rise to the question as to whether the pilots of prehistoric or extraterrestrial flying machines had six fingers on each hand.

An ancient gold figure with four fingers and four toes on hands and feet is exhibited in the patio of the Church of Maria Auxiliadora at Cuenca, Ecuador, together with many other articles of alleged Inca or pre-Inca origin.[6] This god is holding a serpentine staff in the right hand and a rod with the solar disc in the left. His cheeks show traces of dropping tears.

In the centre of the famous Tiahuanaco Calendar Gate stands out the 'Weeping God' whose hands are likewise four-fingered and he also bears two rods.

Why should these primitive sculptors and artists have cast or painted hands with six or four fingers instead of five? It must be borne in mind that the ancient legends of the Americas are solidly based on the tradition of the Sons of the Sun who were different from the men of Earth. After having examined the tracks and finger-prints of astronauts from space who might have explored the planet Earth thousands or millions of years ago, let us now consider the startling possibility that their portraits may still be 'hanging' in some rock gallery.

In 1933 French army lieutenant Brenan, stationed in the Sahara, led a reconnaissance detachment into an unknown canyon in the Tassili Range. On the background of a lunar landscape he suddenly noticed some mural-type rock painting. The officer jumped down from his camel unable to believe his

[4] G. H. S. Bushnell, *Peru*, London, 1963

[5] *Eureka* (annual science digest), Moscow, 1972.

[6] E. von Däniken, *The Gold of the Gods*, London, 1973.

eyes. In that mass of dead rock he saw pictures of a live world—giraffes eating tree leaves, hippopotami splashing in water, elephants running with trunks raised. But duty is duty and Brenan with his troops were forced to leave this prehistoric picture gallery to continue their march. Another 12 kilometres in the burning sun and silence of the desert, and another surprise—more rock paintings.

Four-fingered god on the Gate of the Sun in Tiahuanaco, Bolivia.

Lieutenant Brenan's interest in this newly discovered antediluvian Louvre was shared by his comrade Henri Lhote. Unfortunately they had neither the leisure nor the means to do anything constructive about this archaeological discovery. Then the Second World War broke out and it was only in 1956 that their long-delayed expedition, consisting of fourteen people and thirty camels, reached Tassili. The rocks were covered with thousands of skilfully drawn sketches in twelve or thirteen layers, one superimposed on another. It is apparent from these pictures, and the extinct animals portrayed, that

8,000 or 10,000 years ago this region of the Sahara was a fertile land where people hunted and lived happily. History itself is recorded on those rocks. Female figures with the heads of birds even suggest a possible link with Egypt 5,000 or 6,000 years ago.

At a place named Jabbaren, Professor Lhote found paintings of prehistoric sheep together with other figures known as *round-heads*. These paintings are at least 9,000 years old. The

Prehistoric rock frescoes of Tassili in the Sahara appear to be portraits of astronauts.

figures had enormous round heads, strange collars, two eyes, but neither mouth nor nose. The images were so strange that the members of the expedition jokingly called them the *Martians*. In the Tuareg language *Jabbaren* means the *Ravine of the Giants*. And giants indeed they must have been, for the height of some of these *Martians* was 6 metres, and the strange resemblance between these Jabbaren Titans and present-day astronauts in space-suits and helmets is very striking.

By a stretch of the imagination a sequence of events that could have occurred, unfolds. While swift archers hunted antelope on the green savanna, a spacecraft with a group of astronauts landed. At first the hunters scattered in fear but slowly, as curiosity overcame them, one by one they emerged from behind bushes and rocks. The natives were sure that the astronauts were gods because they had come from the sky. Then little by little, the spacemen approached them, finally made friends with them and learned their way of life. They held in their hands the bows and arrows of both the black and the white hunters who were probably most amused by their imitation. The cosmic explorers studied the buffalo and antelope. When the mission was completed the spaceship took off in a burst of flame setting fire to the grass. Completely bewildered by the spectacular departure, the archers watched the disappearance of the sky-gods into the blue vault of the heavens. The tribal artists immediately drew portraits on rock of those celestial Titans so fresh in their mind's eye.

As the highly artistic drawings of animals show great accuracy in proportion and detail, it is only logical to suppose that the original portraits of the *Martians* of Jabbaren were equally a correct resemblance. Those visitors from the stars may also have left gifts with the simple hunters of the savanna. Metallic or mineral souvenirs may still exist, buried deep in the sands of the Sahara Desert with the bones of the tribal chiefs. Some may have even been discovered but subsequently classified as Egyptian or Carthagenian.

About 100 years before Brenan's discovery, a British lieutenant, George Grey, also accidentally found a most remarkable picture gallery of prehistoric art. Australian anthropologists are amazed to this day that this young English officer, sailing all the way from England, landed at the mouth of the Prince Regent River, marched an exploration party of twelve men through what is still one of the most inaccessible parts of Australia, to the river's source, and found the Magic Caves he did not even know existed. In the hottest and rainiest season of the year this was not an easy exploit. The sandstone gorges of the Prince Regent Valley comprise what is probably the worst travel in Australia. Rain fell in sheets; floods poured down the ravines in fierce cascades. The party lost sheep, ponies and supplies. Neither Lieutenant Grey nor his second in command, Lieutenant Lushington, nor Mr Walker, a naturalist, Mr Powell, a surgeon, nor any of the other men had any idea that they were trespassing the forbidden territory of the Magic Caves of the Kimberleys.

The aboriginal guardians of this sacred district, armed with spears, attacked the unfortunate explorers because this was the domain of the *Rainbow Serpent* which age-old tradition bound them to defend from trespassers who might violate their ancient cult. Three spears hit Lieutenant Grey who fell to the ground. But he managed to shoot dead the leader of the attacking party which, faced with the gun magic of the invaders, immediately retreated. Despite his serious wounds Grey pushed onwards, and after a two months' march, covering only about 115 kilometres, the British expedition finally reached Mount Hann, the crest of the Kimberley Plateau. It was there that the Magic Caves were discovered by pure accident.

The cave rock paintings depict figures with haloes, a mysterious inscription which has never been deciphered, and bird-men which are almost a replica of the bird-men of Easter Island or the bird-women of Tassili. These rock paintings belong to the cult of the sky-god named Wandjina and to the Rainbow Serpent. The Kimberley rock paintings are completely different from all Australian aboriginal art, though attempts to imitate them have been made elsewhere in the land.

In no other part of Australia were so many colours used—red, yellow, black, white and blue. Blue was used nowhere else on the whole continent. The quality of Wandjina artwork is of a superior class with lines giving depth and shading, making it three-dimensional. The Australian aborigines do not pretend to be the painters of these strange pictures. All they claim is that in order to preserve the sacred images from deterioration they have retouched them over the centuries. The originals were painted in the *Dreamtime*, that forgotten era of man's infancy, the time of the first men and the sky-beings.

These Wandjina figures have puzzled scholars for a long time. Australian researcher A. W. Creig, writing in a magazine in 1909 said: 'Not only is the artistic enthusiasm of these savages beyond all reasonable expectation but it finds its expression in the delineation of an object which is altogether beyond the range of their every-day experience, and which would appear from its constant repetition to be of a symbolical nature—a mouthless, halo-crowned figure, clothed in garments which could not have been familiar to the nearly-naked artists who depicted it.'[7]

The *Mouthless Ones* are painted in bold lines on a white background. The head consists of an oval band, either in red or yellow with black outlines. The eyes are shown in black, the nose is rarely shown and the mouth, never. If the portraits are full-size, the models must have been giants 5 metres tall. The *Martians* of Jabbaren are 6 metres tall and they, too, have no mouths. What is bizarre about the Wandjina figures is their fingers and toes which vary in number from three to seven. Some of the bodies appear to be sexless without genitals or clothing, resembling again some of the *Round Heads* of the Sahara.

It must be realised that many of the Wandjina paintings have faded considerably. This indicates great antiquity which is supported by a number of other facts—for instance, a dark rust-coloured glaze covers engravings and rocks alike, the walls on which the paintings had been done show signs of erosion and there is even geological faulting in some places. It has also been established beyond doubt that the aborigines

[7] *The Lone Hand*, 1 May, 1909.

already existed at the time of the extinct bear-like *Diprotodon* as well as that of the *Panaramittee crocodile* that lived in former lakes and rivers of the Central Australian desert. The rock paintings of these animals consequently prove that the aborigine was already painting one million years ago!

There is no way of dating the rock painting of the *Mouthless Ones* in the caves of the Kimberleys. No charcoal was used nor ash left on the ground to permit a carbon-14 test. The Wandjina paintings are supposed to be portraits of sky-gods with rainbow haloes. Could the Rainbow Serpent have been a long spacecraft with a colourful display of lights? Are they representations of beings in space-suits delineated by the fearful yet keen and receptive eyes of Australia's primitive dwellers? Australian tribes attribute their cultural heritage to the heavenly messengers—Baiame, Daramulun and Bunjil. One myth speaks of the sudden arrival of Guriguda, a female deity whose body was covered with quartz crystal instead of skin, flashing lights in all directions. Perhaps she was just an astronaut in a scintillating plastic space suit.

In quest of other traces left by the galactic civilisers—a glance at the Tigris–Euphrates Valley. Among the clay tablets and stone carvings of ancient Babylon there is a weird image of a human head inside a fish's head with the body covered with scales. There is considerable speculation in connection with this engraving. Perhaps still another representation of an astronaut in a space-suit. The legend of Oannes, the being depicted in the engraving, seems to confirm this conjecture. Berosus, a Babylonian priest who lived at the time of Alexander the Great, wrote an historical account of Oannes:

> At Babylon there was in these times a great resort of people of various nations who inhabited Chaldea, and lived in a lawless manner like the beasts of the field. In the first year there appeared, from that part of the Erythraean Sea which

> borders upon Babylonia, an animal destitute of reason by name of Oannes, whose whole body (according to the account of Apollodorus) was that of a fish; that under the fish's head he had another head, with feet also below, similar to those of a man, subjoined to the fish's tail. His voice, too, and language, was articulate and human; and a representation of him is preserved even to this day. This being was accustomed to pass the day among men but took no food at that season, and he gave them an insight into letters and sciences, and arts of every kind. He taught them to construct cities, to found temples, to compile laws, and explained to them the principles of geometrical knowledge. He made them distinguish the seeds of the earth, and shewed them how to collect the fruits; in short, he instructed them in everything which could tend to soften manners and humanise their lives. From that time nothing material has been added by way of improvement to his instructions.[8]

Probably this 'super-dolphin' Oannes, who taught mathematics and science to the ancestors of the Sumerians, was considered 'destitute of reason' because communication between this creature and primitive inhabitants of Mesopotamia was difficult, since Oannes knew so much and they so little. The Babylonian priest added that 'after this there appeared other animals like Oannes'.

This mysterious chronicle becomes clearer in the light of our theory of cosmic civilisers. Clad in space-suits they landed in the sea somewhat in the manner of an Apollo splash-down in the Pacific, and then came ashore to carry out their mission. The culture of Oannes was ready-made and complete—it did not have to be improved on but merely developed. And develop it did. The savages of Mesopotamia began building ziggurats, the skyscrapers of antiquity, and irrigation canals. The priests

[8] I. P. Cory, *Ancient Fragments*, London, 1832.

busied themselves solving problems in higher mathematics, plotting the course of the planets, predicting eclipses and charting the heavens. The Oannes's cosmic mission paid off dividends that we are still collecting in the twentieth century.

From Babylon, with its colossal ziggurats which served as temples and observatories for astronomer-priests, over the Mediterranean and Atlantic to a Mexican 'ziggurat'—the Temple of Inscriptions at Palenque, Yucatan.

This tiered pyramid, with a temple on top, has a long stairway. There are other edifices at Palenque such as the Temple of the Sun, the Temple of the Cross and the Grand Palace, one and all full of remarkable carvings and inscriptions. One inscription says that on the 6th Kaban 10th mol (2 September, 503 A.D.) a convention of all the astronomers of the Mayan Empire took place there. This discovery that makes of Palenque the scientific centre of Mexico, is very important. Some of the astronomical figures recorded there are incredibly accurate. Our astronomers equipped with most intricate instruments calculate the duration of the lunar month to be 29·53059 days. The Mayan calendar shows the duration of the lunar month as 29·53086 days. It is baffling that the Mayan priests should have arrived at such an accurate figure.

In 1948 Dr Albert Ruz-Lhuiller of the Mexican Institute of Anthropology began his three-year excavation work on this pyramid which, although discovered last century, escaped a thorough examination by the archaeologists. Central American pyramids served as pedestals for small temples which sat on their pinnacles. Unexpectedly, the Temple of Inscriptions contained a tomb. Deep at the bottom of the pyramid Dr Ruz found a sarcophagus containing the skeleton of a man in a jade mask together with a collection of jewellery. The skeleton was presumed to be that of a high priest or a king. In one hand he held a small cube and in the other a sphere. The date of the burial was A.D. 633, according to our chronology.

The engraving on the stone lid of the tomb presented a riddle, but riddles are not unusual at Palenque as exemplified by the perfect Christian cross found in one of the temples that existed centuries before the Catholic conquistadores. The bas-relief on the sarcophagus of the Maya priest represents a male figure bent forward with hands resting on some handles or knobs. He appears to be inside a cylindrical cabin of a flying craft with a fiery exhaust. One can almost perceive a control panel in front of the man as well as a nose cone. The inscriptions on the border indicate the signs of the sun, moon and constellations.

Probably the most striking detail of the carving is the manner in which the hands are placed on protruding parts of the inside of the vehicle which might be taken for levers or handles. A close-up of this portion of the carving will demonstrate this more successfully than any verbal description. It is even more curious in that ancient Mexico had no vehicles of any kind. It may be rash to claim this as a representation of a prehistoric aircraft or spaceship. But if Palenque was the Greenwich of the Mayan Empire, frequented by the most competent astronomer-priests, who possessed secret knowledge reserved only for the members of their caste, it would have been only natural for the colleagues of the deceased high priest to erect a suitable monument to him embellished with a cryptogram of ancient science. Life on other planets and visitations by spacemen in the past was part of that arcane tradition.

That the age of Central American civilisation may be much older than accepted at present, can at least be seen from the Mayan date carved on the calendar stone of Tikal, which is 12,042 B.C. in our chronology.[9]

This enigma might have been solved had not the over-zealous Bishop of Yucatan Diego de Landa burned 224 priceless Mayan manuscripts. Had it not been for European

[9] N. Zhirov, *Atlantis*, Moscow, 1970.

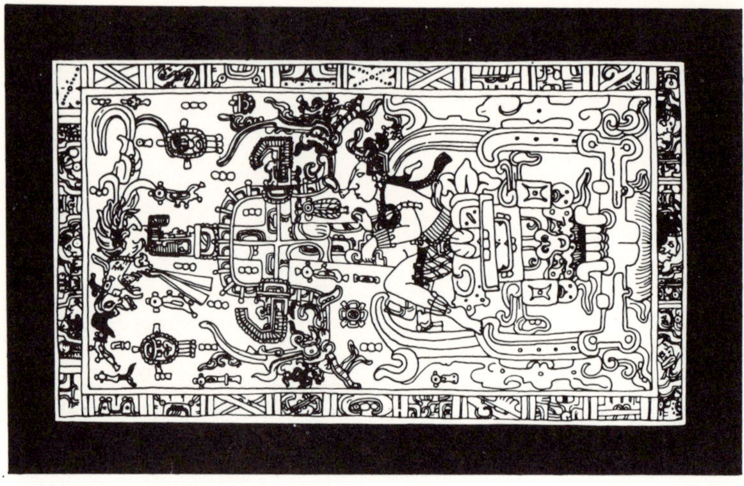

Archaic aircraft or spacecraft with pilot's hands on controls? (Maya tomb carving, Temple of Inscriptions, Palenque, Mexico.)

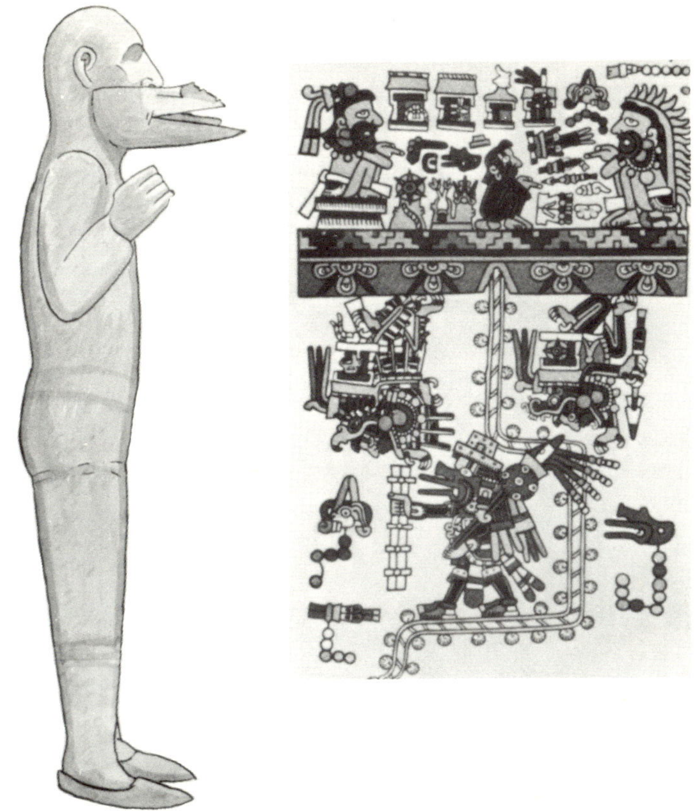

Quetzalcoatl, Messiah of Mexico, shown without his 'wind mask' when in Heaven, but with one when on Earth (Vienna codex). This mask could be easily detached as his statuette in Toluca Museum, Mexico, demonstrates. Breathing aid of a space visitor?

vandals of Landa's type, knowledge in Mayan history would have enriched the world. But all that remains today of that great civilisation are 200 pure-blooded Mayas and three manuscripts.

The Central American culture-bearer Quetzalcoatl, the Plumed Serpent, is known to have come to Mexico in an airship. He wore a tunic, his skin was white and he was bearded. Quetzalcoatl is often portrayed with a 'wind mask' in the shape of a long snout or beak. His statuette in the Toluca museum indicates that this mask could be easily detached. The significant thing is that when shown in 'heaven' with the gods—in such codices, for example, as the *Vienna*, Quetzalcoatl did not wear this mask. It is only when he came down to earth that he

Quetzalcoatl with his 'wind mask' inside jaguar's head might be a Maya artist's idea of a space helmet. (Nochislan, Mixtec culture.)

put it on. It may not be too unreasonable to conclude that Quetzalcoatl's breathing aid was worn only in an atmosphere that was different from that of his own home planet.

Certain sculptures depict Quetzalcoatl's face inside the mouth of a snake, reminiscent of Oannes and the fish's head. This may be a symbolical representation of the space helmet possibly worn by the cosmic civiliser. The same can be said about Quetzalcoatl's image on a vase in Nochislan classified as belonging to Mixtec culture. The stellar civiliser's face is hidden within a jaguar's head but he wears his 'wind mask' as well. There are some beautiful paintings of Quetzalcoatl in the Palacio Nacional in Mexico City. On one mural he is shown descending from the sun in a serpent-like airship, bringing all the benefits of civilisation to the waiting inhabitants of Mexico. The Republic of Mexico and the Empire of Japan are the only countries in the world which recognise the cult of a celestial civiliser.

When enemies disrupted his cultural mission, Quetzalcoatl departed on a raft made of serpents. Another version of his departure speaks of his entering a funeral pyre and disappearing in a pillar of fire, blazing up towards the planet Venus. The Maya–Aztec tradition perpetuated Quetzalcoatl's promise to return. Accordingly, when Cortés landed in Mexico with his band of adventurers, he was mistaken for Quetzalcoatl. This unfortunate error resulted in the downfall of the Aztec Empire.

Can the Quetzalcoatl myth be interpreted as the landing of a celestial torch-bearer of culture, in a long craft, wearing a mask before becoming acclimatised to the new environment? The success of this cosmic mission is proven by the superior quality of Mayan astronomy, the calendrical system and the colossal pyramids which proudly stand in Mexico today.

The Aztec priests alleged that before departing skyward Quetzalcoatl buried a treasure in the earth. So far, no one has ever attempted to search for it. It is probably more precious than gold or jade as it, no doubt, consists of manuscripts and artefacts disclosing the secret science left by the messenger from space. It could be that deep under the tomb of the Maya priest of Palenque lies a cache of artefacts hidden ages ago. In line with this theory, perhaps the cube and sphere in the hands of the skeleton may have some bearing on this mystery.

It may be only a coincidence but the Khufu Pyramid in Egypt embodies in its geometrical proportions the solution of a mathematical problem that has always seemed impossible—the squaring of the circle.* The cube and sphere in the hands of the Maya priest entombed at Palenque also become a square and a circle when projected in two dimensions. This would even link it to the astronaut bas-relief on the lid of the tomb. For orientation in space a flight indicator—a sphere with a cube inside—provides the angles of declination as well as the three axes of space—breadth, length and height.

Do the stone giants of Easter Island, with their sad, narrow eyes looking into infinity, hold a secret, too? In his book about Easter Island, French author Francis Mazière poses the same query. An old native told him a tale which is more than curious. 'Only few stars are inhabited; the dwellers of Jupiter have established communication between planets. The human body can withstand the conditions of other planets for no longer than two months. There is a planet without earth or plants composed only of water and stone.'[10] What is astonishing about this folklore is that it treats of space in a scientific manner. Did the original Easter Islanders, now almost extinct, know the mystery of the stone statues?

Among archaeological ruins, rock paintings and exhibits in museums lurk souvenirs left by visitors from space. When those artefacts are finally identified, we should have our long-awaited evidence of former contacts between solar systems. Meanwhile, man must bear in mind that he is not unique in the unimaginable expanses of our galaxy.

* The Khufu Pyramid is so constructed that the perimeter of the base is equal to the circumference drawn with a radius equal to the pyramid's height.

[10] F. Mazière, *Fantastique Ile de Pâques*, Paris, 1965.

18: *Man Faces the Universe*

Civilisation is the dynamic process of the eternal struggle between the past and the future. When new ideas gain strength and momentum, the old must give way. In the past, whenever the barriers of time or space were broken, humanity advanced. Mankind's consciousness expands as each discovery reveals the universe as vaster and older than had been thought.

Scientific revelations from Copernicus to Darwin have indicated man's true place in Space–Time. He is no longer the 'lord of the Earth' on a stationary planet around which the sun and all the stars move. Moreover, he is no longer the only being that possesses intelligence.

Copernicus, whose 500th anniversary was celebrated while this book was being written, revived the opinions of Anaximander and Aristarchus that the Earth is a planet orbiting the sun. This discovery decentralised the Earth in creation and undermined anthropocentrism.

Half a century after Copernicus, Giordano Bruno declared that there were a multitude of inhabited worlds. He was summoned to the Holy Inquisition and charged with heresy. Before his execution in Rome in 1600 he courageously declared: 'Perchance you who pronounce my sentence are in greater fear than I who receive it.'

After studying the works of Lucretius and other classical writers who had propounded views on evolution, Charles Darwin decided to test and develop their ideas. His historic voyage (1831–6) aboard the *Beagle* to the Galapagos Islands provided him with evidence to demonstrate that all living organisms had evolved from a few primordial forms. His greatest achievement was the presentation of a logical progression which showed that life-forms improve with time.

The dominant ideology was annoyed at this scandalous revival of the agnostic philosophy of the ancient Greeks. Bishop Wilberforce shouted anathemas at Huxley who defended Darwinism at Oxford in 1870. During a lecture on evolution, the wife of scientist David Brewster swooned, so upset was she on hearing statements about woman's lowly origin so shockingly close to animals!

This dethronement of the 'lord of Earth' in the nineteenth century contributed largely towards crushing the fallacy of man's aloofness from Nature. When Albert Einstein proposed a new model for the old universe, proving that every reference system has its own time, space–time was demonstrated to be entirely dependent on the observer. This breakthrough was a repetition, on a cosmic scale, of the Copernican challenge to geocentrism. After centuries of intellectual obscuration, we are at last able to envisage our true position in the universe.

Mankind stands on the outer rim of the Milky Way galaxy on a tiny planet completely lost among the 100,000 million suns which comprise our island universe. But as Greg Benford put it:

I am
Not great or small
But
Part of All.[1]

Today man must be made to realise that he can not exist without Nature but that Nature can well do without him. In a century mankind has poisoned the atmosphere, rivers, lakes and the oceans, brutally destroyed animal and plant life in its

[1] G. Benford, *Deeper than Darkness*, New York, 1970.

selfish chase for food, skins or fuel. Some species on this planet are becoming extinct as a result of the slaughter of creatures having as much evolutionary right to develop as man. What is more, the imbalance of the ecological cycle is even endangering man himself.

Man is the only creature on Earth that destroys his own species on a mass scale. In the armed conflicts of the seven decades of the twentieth century, 36 million civilians and 27 million military personnel were killed.[2] The number of wounded, incapacitated and psychologically dislocated was many times greater. Perhaps this lyric by a poet can describe our planetary tragedy more dramatically:

Writhing, and in its own blood drenched
As if its very soul were wrenched,
This planet is a lumplike curse
In the throat of the universe.

With bases thrown up here and there
Armies alerted everywhere,
And filth man has piled up so high,
The earth, ashamed, sobs out a cry.*

Thus wrote the contemporary Soviet poet Eugene Yevtushenko. 'Wars begin in the minds of men,' says the preamble to the constitution of UNESCO. The spectacle of exploding bombs and fiery sprays of napalm is horrendous. But the burning hatred of those who are spurred by nationalistic and/or religious fanaticism is even more ghastly and all too often has triggered off the holocaust. The world stockpile of nuclear weapons is now equivalent to ten tons of TNT per head

[2] D. Wood, *Conflict in the Twentieth Century*, the Institute for Strategic Studies, London.

* Translated by Andrew Tomas, rendition by Elaine Ackerman.

of Earth's population whereas only a few grammes are needed to kill one individual. It follows that the remainder of the bombs threaten the planet itself. Will the Earth become an asteroid belt, like Phaethon?

Some 20-odd years ago a few thinkers foresaw danger in our present-day trends. Bertrand Russell, addressing the House of Lords in 1958, said that it was possible that no human being would exist by the end of this century, and added: 'It is a race between prejudice and death, and I honestly don't know which is going to win.' He insisted that only World Government could make secure the continuance of the human species.

Einstein held a similar opinion, expressed in a letter written in 1952: 'To me it is enough to know that the continuation of the existence of human beings is in serious doubt if no supranational solution can be achieved.'

Philosophy, religion and ethics are meaningless and useless if they ignore the basic truth that man is a living fragment of a universe of many, many forms. Science has already proven that any dualistic ideology separating man from Nature, animate from inanimate, matter from spirit, is self-deluding. Man's existence on this planet becomes meaningful when considered against the background of a long evolution. Nor is he its end product.

It would be timely to review man's situation with regard to other galactic civilisations and consequences of his fusion with different cosmic cultures. To the average individual, preoccupied with day to day earthbound activities, the idea of contact between worlds in space seems unrealistic. A person in that position may be most astonished to learn that the prospects of such contact are already sufficiently real to both scientists and government agencies, since Space Law has already been established on an international scale.

One of the first attempts to draw up a framework of Space Law was carried out in 1957 by Harvard professor Andrew G.

Haley. This new branch of jurisprudence, which he named *Metalaw*, states plainly that:

> In Metalaw we deal with all frames of existence—with intelligent beings different in kind. We must do unto others as they would have done unto them. We must treat them as they desire to be treated.[3]

The resolution made by the United Nations General Assembly on 20 December, 1961, at its 1085th plenary meeting, states that: 'Outer space and celestial bodies are free for exploration and use by all states in conformity with international law and not subject to national appropriation.' It is commendable and sensible that the nations of the world are not planning to try to institute colonialism on other planets.

In 1962 Soviet jurists proposed a striking clause for UNO Space Treaty: 'The signatories to this Treaty shall regard astronauts as the messengers of mankind in space.'

The Inter-American Bar Association, at their conference in Bogota, Colombia in 1961, adopted the *Magna Carta of Space* as submitted by William Hyman. Part II, *Interplanetary Affairs* reads as follows:

> That in the event it proves there are intelligent beings on any other world, their sovereignty shall be recognised and their laws respected by all peoples of Earth. That aggression or conquest or warfare of any kind shall never be waged by any earthly nation or group of nations against any other inhabited world in the Solar System. That Earth shall carry on a peace policy throughout the universe.

One of the early works on Space Law was written by Myres S. McDougal, Harold D. Lasswell and Ivan A. Vlasic. In this

[3] A. G. Haley, *Harvard Law Record*, Vol. 24, No. 2, 1957.

authoritative book the scholars say that 'the era of astropolitics may bring man in touch with other advanced forms of life'.[4] They theorise that 'an advanced astronautical power will have kept the Earth under surveillance for a long time'. These legal experts foresee three different types of contact with inhabitants of other worlds, and propose suitable policies in each case.

Firstly, man meets an inferior civilisation, and uses minimum interference. Secondly, our astronauts come across a culture on more or less the same level as our own. Experience on Earth offers the solutions. Thirdly, earth civilisation meets a more advanced culture. Here the initiative would lie entirely with the superior power. And these professors speculate that, in view of our sad past history, our Earth may have already been on probation. They raise this question: 'Shall Earth be isolated from the rest of the universe until it has learned to mend its ways and achieved fitness for membership in a larger commonwealth?'

The report of the U.S. Committee on Science and Astronautics, prepared by the Brookings Institution for NASA in 1961, likewise envisages possible contact with other cosmic cultures. It says—'if intelligence of these creatures were sufficiently superior to ours, they would choose to have little if any contact with us'.

Our emergence into space may be likened to the Era of Discovery. The explorations of the New World by Columbus, Magellan, Cortés, Pizzaro and the rest completely changed the course of European civilisation. The Renaissance was the rediscovery of classical culture and science, and we are still reaping the benefits of this historical reversion. These events might be repeated when, with even more dramatic consequences, terrestrial civilisation establishes communication with a superior galactic race that is more advanced in science, technology,

[4] S. McDougal, H. D. Lasswell and I. A. Vlasic, *Law and Public Order in Space*, Yale University, 1963.

ethics, economics and sociology. With the right approach, the advantages of this fusion of galactic civilisations might be colossal.

But what would happen if we were suddenly confronted today with an advanced galactic civilisation? To an unprepared society a sudden encounter with a cosmic empire could spell a temporary disaster to world economy resting on the stock market which fluctuates with current affairs. For example, the possibility of higher and cheaper energies looming on the horizon, whose existence can be conjectured from the many photographs of extraterrestrial spaceships, all power companies —oil, gas, electricity—would have the bottom knocked out from under them. After the first shock, perhaps the mere presence of a super-power might incite the establishment of a supra-national body able to stabilise the chaotic economic and political situation of the world.

It is certain that all political systems would experience the shock of diminished authority. If a superior celestial civilisation became a cognisable fact, the prestige of all national leaders would shrink. The nation states might then perforce join a planetary federation, and full power could be given to the United Nations Organisation, for instance, to deal with world affairs, using international police where necessary. Such a shifting of power to a world authority should, in the long run, be beneficial to mankind as a whole.

Needless to say, the ethical precepts of any given religion need not be invalidated by a fusion of galactic cultures. However, certain religious dogmas might be affected, particularly that of monotheism. Logic, even today, has noted the disparity between the belief in a personal God and the existence of an infinite impersonal universe of billions of galaxies. It is mathematically impossible for infinity to have only one absolute centre. Einstein has summed up this incongruity: 'The main source of the present-day conflict between the spheres of religion and science lies in this concept of a personal God.'[5]

On the other hand this might be only a generalisation in view of the fact that Allah-worshipping Arabs have not only preserved the scientific heritage of the ancient Greeks but even developed it. Neither did faith in Jehovah prevent the Jews from promoting medicine and chemistry for many centuries.

Yet for some obscure reason the Christian Church has maintained a suspicious attitude towards science and technology, which is typified by the sentence of St Polycarp (second century) uttered upon seeing ordinary Egyptian water clocks for the first time: 'In all these monstrous demons is seen an art hostile to God.' As recently as 1864, the *Syllabus of Errors* plainly said that the Christians were 'forbidden to defend as legitimate conclusions of science' if they were contrary to the doctrine of faith. This complex problem should be solved by theologians and historians rather than by the author whose aim is merely to point to the collision of outdated beliefs with modern science.

We must also take into account that all Christian denominations are based on the Nicene Creed which incorporates the belief in Jesus Christ as 'the only begotten Son of God'. The world 'only' does not seem realistic against the background of the stardust of worlds. Such a concept was appropriate in medieval times when the Earth was thought to be the centre of the universe.

The Buddhists seem to be better prepared for the possible confrontation with a foreign space culture. Professor Reihe Masunaga of Tokyo says: 'In Buddhist history there have been no disputes between religious beliefs and scientific knowledge, and no scientists have been persecuted. This is because Buddhism is not contradictory to science.'[6]

[5] A. Einstein, *Out of My Later Years*, London, 1950.

[6] E. Benz, *Buddhism or Communism?* London, 1966.

A Burmese Buddhist leader, U Chan Htoon, made the following statement in America in 1958: 'Buddhism is in tune with modern cosmology because, in contrast to Christianity, it has from the beginning held out for a multiplicity of worlds. It has always been stressed that our Earth is not the only planet capable of bringing forth and sustaining life, but that it is part of an immense system, of an immeasurable plurality of worlds, and stands in intimate relation to this system.'[7]

Hinduism also appears to be in a safe position, illustrated by this passage in the *Vedas*: 'There is life on other celestial bodies far from the Earth.'

The mental horizons of man are often commensurate with his range of vision. In spite of all the technology now available to man, psychologically he remains geocentric and anthropocentric. This attitude surely will have to give way just as the consciousness of the peasant in the Dark Ages would be totally out of place in this Space Age. Such a mental outlook is incongruous with a science that is capable of studying the furthest reaches of the universe as well as envisaging the fusion of galactic cultures. This science is cosmic whereas the mentality of society is still tribal and sectarian.

Whenever the framework of fact and knowledge is in conflict with the established social and ideological systems, sooner or later a clash compels the outdated form to modify and modernise itself, or else be replaced by a new order. We live in an era of silent, bloodless revolution—the revolt of reason against the ignorance of centuries. This clash has been gaining momentum for the last 400 years, and it is now imperceptibly reaching a climax with the challenge from space.

The planet Earth photographed from space is one integral astronomical body. One of the oldest, most tenacious but now, perhaps, most vulnerable concepts is national sovereignty. Nationalism is nothing more than childish self-admiration on a

[7] E. Benz, *Buddhism or Communism?* London, 1966.

geographical level. It must be extremely comic to see when one arrives from another stellar system.

All the people of this planet are passengers of Spaceship Earth. So let us be friendly companions to make it a really pleasant voyage. In history there have been too many hijackers and we do not want any more.

It is possible to envisage a stable technological civilisation under a planetary administration where, in the course of thousands of years, society would have developed a consciousness of responsibility and altruism. One of the basic faults of modern society is loss of the sense of responsibility and the neglect of duty towards each other as well as towards our home planet. History has proven that an irresponsible population bent exclusively on self-indulgence to the detriment of ideals, must inevitably collapse.

Friedrich Engels said that man was created by work. I would like to add that man can destroy himself by the 'fun-culture' dominating the world today. Without spiritual ideals no progress is possible and a nation that lives for 'bread and shows' vanishes, like the Roman Empire. The patricians were sated with luxury and the mob howled for gladiator combats in the Coliseum. During Nero's reign there were six months of holidays a year, cutting production below survival level. Nero had no concern for the integrity of the Roman Empire nor for the world at large. His famous words—'while I yet live, may fire consume the earth', are still remembered as the acme of irresponsibility and so, ancient Rome perished. Nowadays there are thousands of Neroes loose in the world imitating his psychopathic cruelty, selfishness and perversity. This century is a century of crises—intellectual, moral and political, which must be resolved in one way or another.

A person with a minimum of imagination can see that the accelerating progress of science, particularly in space, will eventually change social patterns on Earth. The next 'revolution' is due to come from the top—from the scientific élite, the intelligentsia, the people who really know. They will use no bombs but the facts of science and no machine guns but the rhetoric of logic.

In 1958, one year after Sputnik I, I circularised a project to bring our economics and politics on a par with the Space Age. The proposal is even more timely today. Some of its slogans were: 'A Planetary Government for the Space Age,' 'A Planetary Pool of natural resources, means of production, manpower and scientific genius,' 'One Planet—one family of nations.'

A family comprises different personalities, mentalities and temperaments yet it is united by kinship. Unity in diversity is not something that is impossible to achieve on a greater scale. The world is already internationalised by political and religious ideologies, finance, commerce, transport, science and technology, press, radio and television, and by other media. The United Nations Organisation could play the role of World Government if it adopted an impartial planetary policy. In the meantime progressive scholars and men of science should form a consulting body in order to facilitate the establishment of a planetary federation in which wars would be made impossible and the problems of pollution, political and economic crises could be solved swiftly and efficiently. The seat of the planetary administration might be located on a space platform orbiting the Earth. This giant satellite with scientific equipment should be able to examine any trouble spot within minutes. There is no reason why this project should be relegated to science fiction—it is constructive and practical.

Is a majority of opinion always a qualification for authority? Knowledge and wisdom are scarce whereas ignorance and mediocrity are plentiful. For centuries people believed in the myth of a flat earth. It is always the few who question the established patterns of thought, all too frequently to their own discomfiture. The man of science, who knows more and errs less, should have a greater say in world affairs. What would happen should we one day establish contact with a civilisation in space, so superior to us that we would feel ourselves galactic provincials? Would the demagogues be able to cope with the situation? Here is where a world assembly of scientists would fit, to represent all humanity, not just Mango-Mango, Sylvania or Nutland. It is always the few who lead the many, and those who have advanced scientific knowledge with the ability to orient themselves in history and the universe, would be best qualified to handle such an emergency.

Sometimes I doubt that we have made any progress since the ancient Greeks in spite of our cars and television sets. 'The world is a mosaic of strange peoples—all of them are human beings,' said Herodotus with tolerance. Man stands on the planet Earth today facing an uncertain future, mainly because he has not set any goals for himself in that future. Progress must have an aim in view.

19: *Reflections and Recollections*

This study began with the philosophical proposition that the atomic core of matter contains the potentiality of energy, life and mind. Everything in the world is interconnected. As Emerson wrote—'a leaf, a drop, a crystal, a moment of time, is related to the whole and partakes of the perfection of the whole'. In this scheme all living beings must be regarded as integral particles of the universe, which includes man. The great planetary society consists not only of humans but also of animals, birds, fishes, insects and plants.

In Australia I once placed a dead beetle near an anthill to see what would happen. It was soon spotted by two scouts that became very agitated, thus attracting the attention of other ants in the vicinity. Within minutes there were dozens of ants running all around, and in less than half an hour, hundreds more were hurrying back and forth, little by little transporting the beetle to their anthill. It was astonishing to see that there were no idle workers and the labour force was just what was required in the circumstances.

In Lucknow, India, one could almost be led to believe that monkeys are capable of reading railway timetables. Even though trains are not very frequent there, curious happenings take place at the station shortly before an arrival. Monkeys, young and old, begin to assemble in greater and greater numbers as the time approaches for the train to pull in. First they run around on the shed roofs and then on the platforms. The more people there are on the platform, the greater their chance of not only scavenging pieces of food from the ground, but also of stealing it from absent-minded children and old people.

When the train leaves, all the monkeys disappear. They are never molested because, as all Indians are aware, the god Hanuman wore a monkey guise.

In the Kulu Valley in the Himalayas I witnessed a touching scene, involving large rhesus monkeys. Manali, a village in that valley, is situated at an elevation of 2,750 metres with towering snowcapped ridges on each side. It was surprising to learn that monkeys can live and survive as far up as the snow line. Once at dusk I was walking along a path in a deodar forest, and in the distance saw a large male monkey sitting on a fence pillar, looking right and left. A procession of monkeys, some carrying babies, was crossing the path on the way to the Beas mountain torrent for a drink. Realising that children with mothers had right of way, I stopped as the monkey cop looked at me menacingly.

In Almora, in the Himalayas, I remember meeting a postman on a mountain road leading into a warm valley. He was carrying a bag of letters and a long staff with a brass trident covered with little bells. I asked him if there were any tigers in the jungle valley deep below.

'Sure, sahib, I have seen many in my life,' he answered.

'But what happens if a tiger is too close for one's liking?'

'I jingle this staff of Shiva and the tiger runs away.'

'What do I do when I meet one, I have no gun?' I asked.

'Good thing—no gun. Tiger's clever, he kills enemy.'

'So you think it is all right for me to go into the valley?'

'Yes, sahib, it is all right to go into the jungle but the Indian Postal Service does not guarantee that you will come out . . . that is another and very serious question, sahib.'

Since I never had any pretensions to being a hero, I changed my plans and returned to the village. Sometimes I am not sure that tigers are not nicer than people. For one thing, a tiger would never eat another tiger, which reminds me of a story I heard aboard a liner in the Arafura Sea as the ship was passing

New Guinea. It was told me by an Australian patrol officer. In the early fifties a native suspect was questioned in connection with the recent disappearance of a man from another village.

'Did you kill him?' asked the officer.

'No kill—only eat,' answered the native lying on a mat, with a skull for pillow, obviously digesting a hearty meal. How to deal with cannibals in the twentieth century who draw a sharp distinction between killing with hate and slaughtering a fat man for the replenishment of the larder with a delicacy?

But we should be tolerant even with cannibals. Evolution is a long-drawn-out affair in which the growth of intelligence becomes apparent only after hundreds of thousands of years. However, evolution assures progress, as the arrow of time points towards the perfection of mind, function and form. Promise of continuous ascent rings in infinite space.

It was one of those glorious tropical nights when the sea is calm and the stars shine like diamonds. There I was as Emerson put it, on 'the lonely earth amid the balls that hurry through the eternal halls'. At those latitudes in the Indian Ocean, bright Canopus and the two Magellanic Clouds could be clearly observed. The atmosphere was crystal clear and the stars of the Southern sky seemed to be aflame and alive. Facing these innumerable solar systems, I pondered upon the plurality of inhabited worlds. The idea that we were alone in this ocean of stars seemed so absurd that I smiled.

I wondered whether Tesla, Marconi or others might actually have picked up signals from one of those distant worlds in space. Unusual radio reception has hardly been studied at all. Visitors from space are found in our museums—these are the stone and nickel–iron meteorites. It is not a coincidence that the ratio of one kind to the other does correspond to the composition of the Earth. If an earth-type planet once revolved in the orbit between Mars and Jupiter, and then disintegrated—these very meteorites would be parts of its remains.

With the discovery of life-building chemicals in space, science is becoming more and more certain that their combinations could have created life in other islands of the starry sea.

The topic of possible visitations of astronauts from other solar systems to our world is fascinating. While no concrete proof of these sojourns exists, there is more than ample indirect evidence to uphold such a conjecture. All ancient civilisations such as Egypt, Babylon, China, India, Mexico or Peru had legends of three distinct periods—first, an era when the gods ruled and introduced all the arts and sciences, followed by centuries of reigns by demigods, and lastly, the period known as solar or divine dynasties consisting of men designated to rule by the gods after their departure.

The strange figures of the giants of Tassili and the Wandjina of the Kimberleys may be considered as being portraits of cosmic explorers until a more convincing explanation is offered. The main difficulty in ascertaining the truth is the extremely brief period of time man has evolved on this earth as compared with the vast scale of geological history.

The concept of cosmic artefacts in hidden caches on Earth is not only an interesting theme for science fiction, but an actual possibility already considered by many serious scientists. The location of these treasure-troves of cosmic souvenirs may not be an easy task. That is why an attempt has been made to seek intelligent patterns in their concealment.

There is hardly any doubt that we shall witness more great space revelations before this century is over. The discovery of even the simplest form of life on another planet would demonstrate the potentiality of life in matter throughout the entire universe, as our small section is materially much the same as any other corner of our galaxy, even 50 million light years away.

We have drawn a picture of the river of life from the minutest organisms to astronomical systems. As a link in the endless chain, man has a definite place in cosmic evolution. He is born with the promise of a higher stage of consciousness. The splendour of his prospects should inspire man to wilfully ascend Jacob's Ladder of Nature leading him to cosmic heights.

Conclusion

After travelling to the lands of Lilliput and Brobdingnag it is time to return to Earth. In one day the whole panorama of cosmic evolution unfolds: in the morning—fishes, birds and animals, during the day—the insects; at sunset—man; and by evening—the living stars. The concept of the inexhaustible atom with potential energy, life and mind in its core, could resolve most controversies in science, philosophy and religion. Truth can never be established until the axioms of the infinity of space and the limitlessness of time are recognised. In a boundless, eternal universe everything is possible—somewhere, some time.

A number of theories promulgated here will probably not be readily accepted, but as Thales once said: 'Time is the greatest judge because it finds out everything.'

If life exists in crystals, planets, insects, animals, human beings and supra-human entities, it may well be that just as life is co-extensive with matter, so may intelligence be co-extensive with life. The question of whether there is life and intelligence throughout the universe will become superfluous once we realise that our planet floats on a boundless ocean of life and intelligence.

Principal Sources

ADABASHEV, I., *Mirovye Zagadki Segodnia* (Riddles of the Universe Today), Moscow, 1969.

AGEL, J., *The Making of Kubrick's 2001*, New York, 1970.

ASIMOV, I., *20th Century Discovery*, New York, 1969.

BRAUN, W. VON, *Space Frontier*, London, 1968.

BENZ, E., *Buddhism or Communism?* London, 1966.

BERGIER, J., *Les Extra-Terrestres dans l'Histoire,* Paris, 1970.

BERNAL, J. D., *The Origin of Life*, London, 1967.

BERRILL, N. J., *Biology in Action*, London, 1970.

BLAVATSKY, H. P., *Secret Doctrine*, Adyar, Madras, 1938.

BLAVATSKY, H. P., *Isis Unveiled*, New York, 1886.

BRANDON, S. G. F., *A Dictionary of Comparative Religion*, London, 1970.

BREASTEAD, J. H., *Dawn of Conscience*, London, 1933.

BRIGHTMAN, E. S., *An Introduction to Philosophy*, New York, 1963.

BROOKINGS INSTITUTION, *Report for NASA*, Washington, D.C., 1961.

BURLAND, C., *Mythology of the Americas*, London, 1970.

BURR, M., *The Insect Legion*, London, 1954.

BUSHNELL, G. H. S., *Peru*, London, 1963.

CALDER, R., *Man and the Cosmos*, London, 1968.

CAPLAN, S. A., *Vnezemnye Tsivilizatsii* (Extraterrestrial Civilisations), Moscow, 1969.

CHARDIN, T. DE, *Let Me Explain*, London, 1970.

CHARDIN, T. DE, *Activation of Energy*, London, 1970.

CHARDIN, T. DE, *The Appearance of Man*, London, 1965.

CHARDIN, T. DE, *The Phenomenon of Man*, London, 1959.

CHARON, J. E., *La Connaissance de l'Univers*, Paris, 1965.

CHAUVIN, R., *Animal Societies*, London, 1971.

CHESTNOV, F., *Odinoki li my vo Vselennoi?* (Are We Alone in the Universe?), Moscow, 1963.

CHILDE, G., *What Happened in History*, London, 1954.

CLEATOR, P. E., *An Introduction to Space Travel*, London, 1961.

CORY, I. P., *Ancient Fragments*, London, 1832.

COTTRELL, L., *The Mountains of Pharaoh*, London, 1955.

DÄNIKEN, E. VON, *The Gold of the Gods*, London, 1973.

DRAPER, J. W., *History of the Conflict between Religion and Science*, London, 1922.

DROSCHER, V. B., *The Magic of the Senses*, London, 1971.

DURANT, W., *The Story of Philosophy*, New York, 1933.

DURRELL, G., *Encounters with Animals*, London, 1971.

EDDINGTON, A., *The Nature of the Physical World*, London, 1964.

ENGELS, F., *Dialectics of Nature*, Moscow, 1940.

FECHNER, G. T., *Religion of a Scientist*, New York, 1946.

FEDINSKY, V., *Meteors*, Moscow, 1959.

FIRSOFF, V. A., *Life, Mind and Galaxies*, London, 1967.

FLAMMARION, C., *La Pluralité des Mondes*, Paris, 1884.

FLEW, A., *An Introduction to Western Philosophy*, London, 1971.

FOX, H. M., *The Personality of Animals*, London, 1940.

FRAZER, J. G., *The Golden Bough*, London, 1907.

GOULD, R. F., *A Concise History of Freemasonry*, London, 1903.

GUTHRIE, W. K. G., *A History of Greek Philosophy*, Cambridge, 1969.

HAGEN, V. W. VON, *Realm of the Incas*, New York, 1957.

HALDANE, J. B. S., *Science and Life*, London, 1968.

HAMILTON, E., *Mythology*, New York, 1953.

HOSPERS, J., *An Introduction to Philosophical Analysis*, London, 1956.

HOYLE, F., *Of Men and Galaxies*, Seattle, 1964.

HOYLE, F., *Frontiers of Astronomy*, London, 1961.

HUGHES, E. R., *Chinese Philosophy in Classical Times*, London, 1966.

HUYGENS, C., *The Celestial Worlds Discovered*, London, 1698.

IONS, V., *Egyptian Mythology*, London, 1968.

JACOT, L., *Universal Evolution*, Geneva, 1967.

JAMES, T. G. H., *Myths and Legends of Ancient Egypt*, London, 1969.

JARMAN, C., *Evolution of Life*, London, 1970.

JEWISH ECYCLOPEDIA, New York, 1925.

JOHNSON, R. C., *Nurslings of Immortality*, London, 1957.

KONDRATOV, A. M., *Tainy Trekh Okeanov* (Mystery of the Three Oceans), Leningrad, 1971.

KONDRATOV, A. M., *Pogibshiye Tzivilizatsii* (Lost Civilisations), Moscow, 1968.

LAPP, R. E., *Man and Space*, London, 1961.

LAROUSSE ENCYCLOPEDIA OF MYTHOLOGY, London, 1960.

LEGAL PROBLEMS IN SPACE EXPLORATION, U.S. Congress, Washington, D.C., 1961.

LEY, W., and BONESTALL, C., *Beyond the Solar System*, London, 1965.

LIFE, *Planets* (Time–Life), 1966.

LOZINA-LOZINSKY, L., *Granitsy Zhizni* (Frontiers of Life), Moscow, 1962.

MCCARTHY, F. D., *Australian Aboriginal Rock Art*, Sydney, 1962.

MCDOUGAL, M. S., LASSWELL, H. D., and VLASIC, I. A., *Law and Public Order in Space*, Yale University Press, 1963.

MARAIS, E. N., *The Soul of the White Ant*, London, 1971.

MARTIN, C.-N., *Le Cosmos et La Vie*, Paris, 1970.

MATHERS, MACGREGOR S. L., *The Kabbalah Unveiled*, London, 1968.

MELLERSH, H. E. L., *From Ape Man to Homer*, London, 1962.

MONOD, J., *Chance and Necessity*, London, 1972.

MORLEY, D. W., *The Ant World*, London, 1953.

MOSS, A. A., *Meteorites*, British Museum, 1971.

MULLER, W. M., *Egyptian Mythology*, London, 1927.

MURRAY, D. S., and JEFFREE, G., *The Anatomy of Man and Other Animals*, London, 1951.

NEWMAN, L. H., *Ants from Close Up*, London, 1968.

OBERTH, H., *Man into Space*, London, 1957.

OUSPENSKY, P. D., *Tertium Organum*, London, 1950.

PACKARD, V., *The Human Side of Animals*, New York, 1950.

PARNOV, E. I., *Na Perekrestke Bezkonechnostei* (At the Crossing of Infinities), Moscow, 1967.

PARRINDER, G., *African Mythology*, London, 1967.

PERELMAN, R. G., *Zvezdnye Korabli* (Starships), Moscow, 1961.

PINCHER, C., *Evolution*, London, 1950.

PONNAMPERUMA, C., *The Origins of Life*, London, 1972.

PRESCOTT, W. H., *The Conquest of Peru*, New York, 1961.

REDDISH, V. C., *Evolution of the Galaxies*, London, 1962.

REEVES, E., *The Anatomy of Peace*, London, 1947.

ROERICH, N., *Heart of Asia*, New York, 1929.

ROERICH, N., *Himalayas—the Abode of Light*, London–Bombay, 1947.

ROERICH, N., *Nerushimoye* (The Indestructible), Riga, 1936.

ROUSSEAU, P., *Man's Conquest of the Stars*, London, 1959.

RUSSELL, B., *History of Western Philosophy*, London, 1971.

SCHRÖDINGER, E., *What is Life?* Cambridge, 1967.

SHAPLEY, H., *The View from a Distant Star*, New York, 1963.

SHAPLEY, H., *Of Stars and Men*, London, 1958.

SHAPLEY, H., *Beyond the Observatory*, New York, 1967.

SHKLOVSKY, I. S., *Vselennaya, Zhizn, Razum* (Life and Intelligence in the Universe), Moscow, 1965.

SIEGEL, F., *Zhizn v Kosmose* (Life in the Universe), Minsk, 1966.

SINNETT, A. P., *Collected Fruits of Occult Teachings*, London, 1920.

SOKOLOVSKY, U., and SHILOV, V., *Fotonny Zvezdolet* (The Photon Starship), Kharkov, 1960.

STENUIT, R., *The Dolphin—Cousin to Man*, London, 1968.

SULLIVAN, W., *We Are Not Alone*, London, 1970.

THOMAS, P., *Epics, Myths and Legends of India*, Bombay, 1961.

VOROBIEV, G. G., *Chto vy znaete o Tektitakh?* (What do you know about Tektites?), Moscow, 1967.

WENTZ, W. Y. EVANS, *Tibet's Great Yogi Milarepa*, Oxford, 1969.

WINDELBAND, W., *A History of Philosophy*, New York, 1958.

WOOD, J. A., *Meteorites and the Origin of Planets*, New York, 1968.

YOUNG, J. Z., and MARGERISON, T., *From Molecule to Man*, London, 1969.

Index